W9-CRB-723

Fifty Years of Neutron Diffraction

The advent of neutron scattering

Fifty Years of Neutron Diffraction

The advent of neutron scattering

edited by

G E Bacon

Adam Hilger, Bristol

Published with the assistance of the
International Union of Crystallography

0 260 - 4425

PHYSICS

©IOP Publishing Ltd and Individual Contributors

British Library Cataloguing in Publication Data

Fifty years of neutron diffraction: the advent of neutron scattering
   1. Neutrons——Diffraction
   I. Bacon, G. E. II. International Union
   of Crystallography
   548′.83   QC793.5.N462
ISBN 0-85274-587-7

Published under the Adam Hilger imprint by IOP Publishing Ltd
Techno House, Redcliffe Way, Bristol BS1 6NX, England

Printed in Great Britain by J W Arrowsmith Ltd, Bristol

Contents

List of Contributors

A F Andresen
Institute for Energy Technology
Box 40
N-2007
Kjeller
Norway

G E Bacon
Department of Physics
University of Sheffield
Sheffield
S3 7RH
UK

J Baruchel
Centre National de la Recherche
    Scientifique
Laboratoire Louis Néel
25 Avenue des Martyrs
166 X
38042 Grenoble Cedex
France

G A Briggs
Institut Laue–Langevin
156 X
38042 Grenoble Cedex
France

B N Brockhouse
Department of Physics
McMaster University
Hamilton
Ontario
Canada L8S 4M1

B Buras
Risø National Laboratory
Postbox 49
DK-4000 Roskilde
Denmark

J W Cable
Oak Ridge National Laboratory
PO Box X
Oak Ridge
Tennessee 37831
USA

G Caglioti
Istituto di Ingegneria Nucleare
CESNEF-Politecnico di Milano
Via Ponzio, 34/3
20133 Milano
Italy

P Chieux
Institut Laue–Langevin
156 X
38042 Grenoble Cedex
France

P Convert
Institut Laue–Langevin
156 X
38042 Grenoble Cedex
France

D Cribier
Departement de Physique Générale
Centre d'Etudes Nucleaires de
    Saclay
91191 Gif-sur-Yvette Cedex
France

H Dachs
Hahn-Meitner Institut für
   Kernforschung
Postfach 390128
D-1000 Berlin (West) 39

P A Egelstaff
Department of Physics
University of Guelph
Guelph
Ontario
Canada N1G 2W1

J E Enderby
Institut Laue–Langevin
156 X
38042 Grenoble Cedex
France

H Fuess
Institut für Kristallographie und
   Mineralogie
Johann Wolfgang Goethe
   Universität
Postfach 11 1932
D-6000 Frankfurt am Main 11
Federal Republic of Germany

J A Goedkoop
Netherlands Energy Research
   Foundation
PO Box 1
1755 ZG Petten NH
The Netherlands

S Hoshino
Institute for Solid State Physics
University of Tokyo
Roppongi
Minato-Ku
Tokyo 106
Japan

P K Iyengar
Bhabha Atomic Research Centre
Trombay
Bombay 400 085
India

B Jacrot
European Molecular Biology
   Laboratory
c/o Institut Laue–Langevin
156 X
38042 Grenoble Cedex
France

W C Koehler (deceased)
Oak Ridge National Laboratory
Post Office Box X
Oak Ridge
Tennessee 37831
USA

G H Lander
Intense Pulsed Neutron Source
IPNS-360
Argonne National Laboratory
Argonne
Illinois 60439
USA

J Leciejewicz
Institute of Nuclear Chemistry and
   Technology
Ul. Dorodna 16
03-195 Warszawa
Poland

H Maier-Leibnitz
Pienzenauerstrasse 110
8000 Munchen 81
Federal Republic of Germany

M H Mueller
Argonne National Laboratory
9700 South Cass Avenue
Argonne
Illinois 60439
USA

F Mezei
Hahn-Meitner Institut für
    Kernforschung
Postfach 390128
D-1000 Berlin (West) 39

R M Moon
Oak Ridge National Laboratory
Post Office Box X
Oak Ridge
Tennessee 37831
USA

R P Ozerov
Mendeleev Institute of Chemical
    Technology
Miusskaja pl. 9
Moscow 125820
USSR

L Passell
Department of Physics
Brookhaven National Laboratory
Upton
Long Island
NY 11973
USA

B M Powell
Atomic Energy of Canada Ltd
Chalk River Nuclear Laboratories
Chalk River
Ontario
Canada K0J 1J0

H Rauch
Atominstitut der Österreichischen
    Universitäten
Schuttelstrasse 115
A-1020 Wien
Austria

G R Ringo
Argonne National Laboratory
9700 South Cass Avenue
Argonne
Illinois 60439
USA

A Yu Rumjantzev
I V Kurchatov Institute of Atomic
    Energy
Ul. Kurchatova 46
Moscow
USSR

T M Sabine
New South Wales Institute of
    Technology
Sydney
NSW 2007
Australia

M Schlenker
Centre National de la Recherche
    Scientifique
Laboratoire Louis Néel
25 Avenue des Martyrs
166 X
38042 Grenoble Cedex
France

C G Shull
Department of Physics
Massachusetts Institute of
    Technology
Cambridge
MA 02139
USA

G L Squires
Cavendish Laboratory
Madingley Road
Cambridge
CB3 0HE
UK

G C Stirling
Rutherford Appleton Laboratory
Chilton
Didcot
Oxon
OX11 0QX
UK

R J Weiss
King's College
University of London
Strand
London
WC2R 2LS
UK

D Wheeler
Institut Laue–Langevin
156 X
38042 Grenoble Cedex
France

B T M Willis
Chemical Crystallography
   Laboratory
9 Parks Road
Oxford
OX1 3PD
UK

and

Atomic Energy Research
   Establishment
Harwell
Didcot
Oxon
OX11 0RA
UK

A D B Woods
Atomic Energy of Canada Ltd
275 Slater Street
Ottawa
Ontario
Canada K1A 1ES

Z Yang
Institute of Atomic Energy
PO Box 275 (30)
Beijing
China

Preface

In 1962 the fiftieth anniversary of the discovery of X-ray diffraction was commemorated by the publication, for the International Union of Crystallography, of *Fifty Years of X-ray Diffraction*, edited by P P Ewald and running to 720 pages. In 1981, a few years after the corresponding anniversary for electrons, a companion volume *Fifty Years of Electron Diffraction* appeared, edited by P Goodman with 440 pages. We now complete the trilogy to celebrate the achievements of 50 years' work with neutrons since the first diffraction experiments of 1936. However, the title *Fifty Years of Neutron Diffraction* is supplemented with a subtitle *The Advent of Neutron Scattering*. This is intended to indicate how the scattering process itself—be it coherent, incoherent, elastic or inelastic—has broadened the concept and its applications far beyond what we normally think of as 'diffraction'.

This volume was conceived and encouraged by the Commission on Neutron Diffraction of the International Union of Crystallography. The Union has contributed financially towards the book's production, with the aim of making its purchase more accessible to individual scientists. With the same aim, the length of the book has had a target of 250 pages. This has posed some problems for myself as Editor. Unable to find space for the early history of the technique in every country and unable to invite contributions and reminiscences from so many worthy writers, I hope that I have retained some friends. I would indeed have liked to include many more articles. However, the aim of the book is not to give a complete survey of the subject—or, even less, a review of the literature—but to try and capture something of its flavour and opportunities and the way in which life for the neutron experimenter has changed during the 50 years.

My thanks go to the individual contributors and to their institutes and laboratories which enabled them to contribute. In particular I would like to thank the Brookhaven National Laboratory who paid for my use of figure 3.4, kindly arranged by Dr L M Corliss. Finally, on a very sad note, I have to record that Dr W C (Wally) Koehler, who wrote the article on 'Neutrons and Magnetism', died while the book was in production. I am very pleased that he was able to make his contribution.

G E Bacon
Guiting Power,
July 1986

1   Introduction: The Pattern of 50 years

G E Bacon

University of Sheffield, UK

Four years after the discovery of the neutron in 1932 it was demonstrated that the neutron could be diffracted. As the Golden Age of physics was slipping away, three crucial papers were published in 1936, the year which also saw the first artificial disintegration of an element.

These three papers are reproduced in the following chapter of this book. In the first paper, reported in March 1936, Elsasser considered the way in which a beam of thermal neutrons, containing a wide range of velocities, would be diffracted by a powder. He showed that there would be a minimum angle, determined by the maximum interplanar spacing of the solid and the peak wavelength of the thermal spectrum, below which no scattering would take place. Three months later, at a meeting of the Academy of Sciences of Paris, Halban and Preiswerk justified this conclusion with an account of a practical experiment using a radium–beryllium source. About a month later the result of a refined experiment was reported in *Physical Review* by Mitchell and Powers. They, too, used a Ra–Be source but diffraction took place from a set of large single crystals, of magnesium oxide, and the diffracted beam was isolated and measured separately, in contrast with Halban and Preiswerk's experiment which measured the sum of the direct and diffracted beams.

Experimentally, the subject now had to await the much more intense beams which were to come from the early nuclear reactors, but there was much progress among theoreticians in the USA. Bloch, in 1936, discussed the magnetic scattering of neutrons, particularly in relation to ferromagnetism, and in 1939 the comprehensive paper of Halpern and Johnson appeared, showing that it should be possible to demonstrate magnetic scattering most clearly with paramagnetic materials. Morton Hamermesh, who worked with both Bloch and Halpern, writes

1

I began my graduate studies at New York University in 1936. I became a theoretical student with Otto Halpern as my adviser, starting in 1938. I also worked at Columbia, doing experimental work (with Cohen, Goldsmith, Manley and Schwinger) and theoretical work with Schwinger (on scattering of neutrons by ortho- and para-hydrogen and deuterium). Halpern and Johnson had written on paramagnetic scattering, and Bloch had worked on polarization by ferromagnetic scattering. Ted Holstein was assigned the problem of depolarization of neutrons by magnetic scattering. I was first asked to develop the diffraction formulas for neutron scattering, which I did very quickly. Then Halpern kept suggesting more topics, and the thesis grew and grew. The part of the thesis I liked best was the derivation of the formula for double refraction, based on a paper of Oseen (1912), which I used to relate the index of refraction to the scattering amplitude. I used this later (1950) to discuss total reflection of neutrons by Fe and Co.

The work was exciting and stimulating. Our professors paid little attention to us. My classmates were H. Primakoff and T. Holstein. Cliff Shull was a little younger. I helped him build an electron Van de Graaff accelerator with which he did the polarization of electrons by double scattering. Most of what I learned, I learned from my classmates and from Julian Schwinger. In 1941, I went to Stanford, where I worked with Bloch and Staub on measuring the degree of polarization in transmission through iron.

The future was transformed by the possibilities made evident by the first 'atomic pile' set up by Fermi and his colleagues in December 1942. This led to the construction of other 'piles'—soon to become known as 'nuclear reactors—and, in particular, the Clinton Pile, a graphite-moderated reactor at Oak Ridge, which operated late in 1943, and CP-3, with a heavy-water moderator, built in the Argonne Forest before this site had been formally designated as the Argonne Laboratory. During 1944 and 1945 these reactors were used with single-axis spectrometers and Fermi choppers for producing monoenergetic neutron beams and the collection of nuclear cross section data. In July–September 1944 rocking curves for a lithium fluoride crystal and measurements of the reactor spectrum were carried out by Zinn and subsequently announced in *Physical Review* in 1946 and reported more fully in 1947. Two early photographs relate to this period. Figure 1.1 shows E Fermi with the electronics for what was probably the first Fermi chopper: figure 1.2 shows W H Zinn at the CP-3 reactor with the crystal spectrometer which he designed in 1944 and which is now preserved as a museum piece at the Argonne National Laboratory. The paper by Zinn (1947) was accompanied by two others, one by Sturm (1947) which gave many more experimental details and scattering data for many elements and a most important paper by Fermi and Marshall (1947). The latter may be regarded as the first revelation to physicists and chemists of the prospects of using neutrons for studying solids. This paper gave

Figure 1.1. E Fermi, around 1947, with what is probably the electronics for the first Fermi chopper.

scattering lengths for some 22 nuclear species, made clear how the scattering amplitude of a nuclear species can be determined and how this amplitude may vary for different isotopes and depend on nuclear spin. It also established by mirror reflection that there is normally a phase change of 180° when neutrons are scattered by a nucleus, but for a few exceptional elements—Li, Mn, H—the phase change is zero.

By 1946 Oak Ridge was establishing itself as the main centre of development. E O Wollan, with his background of X-ray scattering studies of gases, decided to try and avoid the difficulties which had been experienced in interpreting single-crystal reflections, in the face of secondary extinction, by looking at powder diffraction. He found there was indeed adequate intensity and this move to powders was eminently productive. In early 1946 Wollan (with R B Sawyer) had obtained a powder diffraction pattern of NaCl and patterns for liquid H_2D and D_2O. These were impressive enough to attract C G Shull, who joined Wollan in June 1946.

Figure 1.2. W H Zinn at the Argonne National Laboratory with the first crystal spectrometer for neutrons which he designed in 1944 (see Zinn 1947).

The following few years were extraordinarily fruitful. As seen from afar, they culminated in two essential achievements. First, 1951 saw the appearance of Shull and Wollan's classical paper which, among other things, listed the coherent scattering amplitudes for about sixty elements and isotopes. This work really established the potential of neutron diffraction as a technique and tool for studying solids. Henceforth it became commonplace to speak of the 'applications' of neutron diffraction. Secondly, another milestone was the paper of Shull and Smart in 1949 revealing the magnetic structure of manganous oxide, a practical demonstration of the magnetic scattering foretold by Bloch and Halpern and Johnson and a complete justification of the ideas of Néel. It was quickly followed by two other comprehensive papers from Oak Ridge on antiferromagnetic and ferromagnetic materials (Shull, Strauser and Wollan 1951, Shull, Wollan and Koehler 1951). These papers greatly extended the range of interest in neutrons to include researchers in both experimental and theoretical magnetism.

To the rest of the world, struggling to get started in the new technique, this immense burst of activity was both challenging and a little disheartening. The new entrants had the tantalising choice between filling in some of the obvious gaps in the field as explored so far and branching out in more speculative directions which might be slow to produce results. Chalk River was the first establishment outside the USA to have a reactor, and by 1949 Hurst and his co-workers were reporting neutron diffraction from gases. Harwell, and Levy's chemistry group at Oak Ridge, returned to the abandoned battle with the single crystal: secondary extinction was never fully mastered, but it was largely understood. KH_2PO_4 was studied in three separate laboratories and the power of neutrons to depict hydrogen bonds impressed the chemists. Neutron crystallography had arrived. Voices crying in the wilderness suggested that there might even be a future for neutrons in biology.

In the 1950s nuclear reactors with the means of extracting neutron beams to study materials spread across Europe and to India, Australia and the Far East. The early stories of some of these research groups are given in Chapter 4. Meanwhile, improvements were taking place in the reactors themselves and in the efficiency of extracting and using the neutron beams. The early reactors had a neutron flux of about 10^{12} neutrons $cm^{-2} s^{-1}$ and this value has been increased roughly by a factor of ten each decade. The most conclusive and lasting result of the first increases was to foster interest in the inelastic scattering of neutrons, thus taking advantage of the fact that not only did the wavevector of the neutron match the scale of interatomic distances and forces in solids and liquids but, also, its energy matched the scale of vibrations of all kinds—atomic, molecular, magnetic. A theoretical paper by Weinstock (1944) had been followed by Cassels, Squires and others, and in 1951 Egelstaff had directly demonstrated the gain in energy when long-wavelength ($5.7\,\text{Å}$) neutrons were scattered inelastically. By 1954, notably through the efforts of Brockhouse at Chalk River and Jacrot at Saclay, slow-neutron spectrometry was an accepted technique. The first triple-axis spectrometer appeared and figure 7.9, a photograph taken in 1958, shows B N Brockhouse with his first properly engineered version. The flux of the NRU reactor at which the spectrometer is mounted is $3 \times 10^{14}\, cm^{-2} s^{-1}$ and by now the 'constant Q' method has appeared. The study of 'inelastic scattering' has become a recognised discipline and 1960 saw, at Vienna, the first of a succession of conferences on the 'Inelastic Scattering of Neutrons in Solids and Liquids', convened by the International Atomic Energy Agency. Fifty papers on theoretical and experimental contributions to the subject were presented at this first conference. Three years later, at the meeting of the Sixth General Assembly in Rome, the International Union of Crystallography prepared to set up a permanent Commission on Neutron Diffraction, to foster progress and development of the subject. It was evident that the original 1936 diffraction experiment had developed into the vastly wider concept of neutron *scattering* and was now a well established branch of international science.

At this date, as we pass the half-way point in our story, there were about forty laboratories in the world where neutron scattering experiments were being actively carried out. The achievements of inelastic scattering studies had immensely extended the range of interest among the participants, to cover most of solid-state physics and solid-state chemistry, and the advent of reactors giving higher flux would soon permit a crossing of the threshold into biology. These were the days of manganese and chromium, the recognition of non-commensurate magnetic structures and Koehler's un-ravelling of the fascinating structures of the rare-earth metals.

It became possible to contemplate the construction of new reactors at in-dividual universities or, if daunted by contemplation of the problems of reactor operation and maintenance, to plan the 'leasing' of reactors, or portions of reactors, to university researchers. In this way the Science Research Council in the United Kingdom acquired partial use of reactors at the Atomic Energy Establishments at Harwell and Aldermaston for university workers, and joined in providing new, much more advanced, apparatus for neutron scattering. This greatly encouraged the spread of the technique among 'partial' users who, while not considering neutron scatter-ing to be their primary interest, realised that they had some problem which neutrons could probably solve. The administration of the system provided for the instruction and guidance of these temporary users by the laboratory staff and, indeed, depended very substantially on the tact and goodwill of the latter. A similar system operated from the outset at Lucas Heights in Australia.

However, in retrospect, it is evident that a watershed had now been reached, and for several different reasons. With the continuing need for, and indeed the provision of, steadily increasing fluxes it was clear that the needs of neutron scattering were not met by a share in a 'general purpose' nuclear reactor which had been designed for isotope production, irradia-tion experiments and the testing of engineering components intended for use in power reactors. To obtain higher fluxes, specially designed reactors with smaller cores were needed and in 1965 the HFBR reactor at Brookhaven, with a flux of 10^{15}, went into operation. Very soon afterwards a reactor for isotope production, but with approximately the same flux, was working at Oak Ridge. In 1972 a specially designed reactor, again with a flux of about 10^{15}, was in operation at Grenoble in France. The latter, built jointly by France and Germany, was unique in that it was constructed at a new institute, the Institut Laue–Langevin, founded for the exploitation of research using neutron beams. Technically this reactor was unique in having guide tubes which, with much advantage to the experimenters, channelled many of the neutrons away to an adjoining large hall. In the following year, 1973, primarily after much persuasion by E W J Mitchell, the United Kingdom bought a one-third share of the Institute and became an equal partner with France and Germany. Over the next decade the Institut Laue-Langevin grew to be almost a synonym for neutron scattering to a large proportion of the neutron community. More and more of them became familiar with the journey to Grenoble, the CENG canteen and the long walk down the Avenue des Martyrs in the small hours, past the barking dog at the Total garage. The camaraderie helped to maintain a closely knit community, following the tradition established with X-rays from the beginnings in 1912. It was encouraged by a succession of international conferences, which contributed both to scientific progress and international understanding. Figure 1.3 recalls a gathering at Krakow in 1978. A formal group photograph taken at Harwell in 1968 is figure 1.4.

At the same time as increased fluxes had led to the appearance of the 'dedicated' reactor, developments in computer technology and methods of apparatus control had revolutionised the practical manner of carrying out neutron experiments. The changes were further accelerated by advances in position-sensitive detectors and the production of monochromators. Each

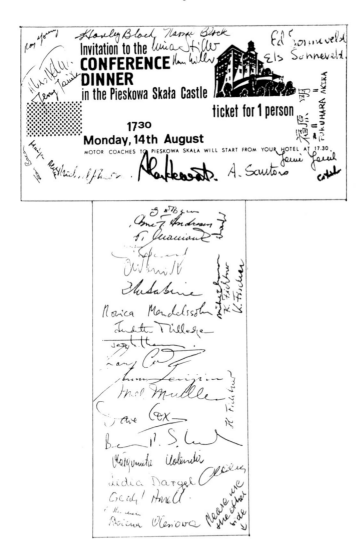

Figure 1.3. A memento of a neutron scattering meeting in Krakow, 1978.

Figure 1.4. A meeting on neutron diffraction at Harwell 1968.

Back Row—Left to Right: 1. D P Treble. 2. J Yerkess. 3. S J Campbell. 4. S J Wright. 5. S Wild. 6. W Roberts. 7. J A Veltze. 8. F Rossitto. 9. D H Day. 10. J Trotter. 11. R Harris. 12. D W Jones. 13. B C Tofield. 14. A Jacobson. 15. F K Larsen. 16. A R Burne. 17. R Jude. 18. M Roudaut. 19. G S Pawley. 20. A Whitaker. 21. S K Sikka. 22. J Osterlof. *Centre Row—Left to Right:* 1. B P Schoenborn. 2. K D Rouse. 3. P Jones. 4. M Rogers. 5. R J R Miller. 6. W C H Alston. 7. J H Katz. 8. D J Winfield. 9. F R Thornley. 10. Z Barnea. 11. F Zigan. 12. M J Yessik. 13. J Moss. 14. A M Afif. 15. J O Burgman. 16. G W Cox. 17. J R Ravenhill. 18. A C MacDonald. 19. P J Webster. 20. J G Booth. 21. J S Plant. *Front Row—Left to Right:* 1. Miss P Papamantelos. 2. Miss J A K Duckworth. 3. Miss E A Yeats. 4. R A Young. 5. B Dawson. 6. C Johnson. 7. G Caglioti. 8. P Coppens. 9. S C Abrahams. 10 B T M Willis. 11. C A Coulson. 12. P P Ewald. 13. G E Bacon. 14. Miss P J Brown. 15. M J Cooper. 16. K C Turberfield. 17. A F Andresen. 18. J C Speakman. 19. Miss J L Cox. 20. Miss P Welford. 21. Miss H Gamari-Seale.

change contributed to the speed at which experiments could be done, to the range and elegance of the problems which could be contemplated and to the sophistication—and inevitably the cost—of all the spectrometers, diffractometers and their associated equipment. An experiment which in 1950 was impossible, or which might have been carried out rather unsatisfactorily in a week, can now be successfully done in 10 minutes. It has become more and more essential to use all the apparatus efficiently for 24 hours a day, so that the administration and planning of experiments become as important as carrying them out. Regretfully in many ways, the role of the small establishment is disappearing and, increasingly, the race is only for the strong.

The failing position of the small organisation in neutron scattering has deteriorated even more because of a further, rather different, technical development. Consideration had often been given to producing alternative sources of neutron beams to replace the nuclear reactor, bearing in mind a likely limit to the flux which can be achieved with a reactor. In the late 1960s some progress was made using pulsed neutron beams generated from electron linear accelerators. After being slowed down to thermal energy the 'white' beam was diffracted by a sample and analysed, at a single angle of scattering, by the time-of-flight method. The technique for doing this was already known from the work of Buras, particularly with the pulsed reactor at Dubna, which is described in section 8.1. In later years, much more intense neutron beams have been produced from spallation sources in which heavy-element nuclei are bombarded by high energy protons. Several sources of this type are now in operation in the USA, Japan and the UK, and others are planned. These spallation sources are described in Chapter 9 and are now being accepted as an alternative, and a rival, to the nuclear reactor. However, the future development of reactors is by no means impossible and a final section of Chapter 9 describes a proposal for a new reactor at Oak Ridge to be used for neutron scattering, isotope production and irradiation experiments. The flux would be at least 5×10^{15} and the neutron beams would outshine the present Oak Ridge reactor by ten times and the Institut Laue–Langevin by four times.

It seems certain that the future of neutron scattering experiments is assured well beyond this period of 50 years which has now passed by. Most likely the work will be done at a small number of international institutes. The past few years have seen several international agreements for the use and construction of expensive neutron-beam equipment. For example, there have been agreements between the USA and Japan concerning instruments at Brookhaven and Oak Ridge and between the UK and both India and West Germany for the exploitation of ISIS, the new spallation source at the Rutherford Appleton Laboratory. Centralisation is, however, not without disadvantages: it is more difficult to plan a programme and less easy to train research students, who in the 1960s and 1970s could easily be introduced to neutrons informally at a 'local' reactor. It is not easy to foretell the precise activity of future experimenters and their samples. Having begun with a stop-watch and a turning-handle, in the genuine 'string and sealing wax' tradition of the neutron's discoverers, we have already advanced a long way in reaching the sophisticated experimental apparatus of today. A stage has now been reached when the time occupied by the journey to and from the neutron source may be greater than that taken up by the experiment itself. What next?

Twelve Classical Papers

Egelstaff P A 1951 *Nature* **168** 290
Elsasser W M 1936 *C. R. Acad. Sci., Paris* **202** 1029
Fermi E and Marshall L 1947 *Phys. Rev.* **71** 666
Halban H and Preiswerk P 1936 *C. R. Acad. Sci., Paris* **203** 73
Halpern O and Johnson M H 1939 *Phys. Rev.* **55** 898
Mitchell D P and Powers P N 1936 *Phys. Rev.* **50** 486
Shull C G and Smart J S 1949 *Phys. Rev.* **76** 1256
Shull C G, Strauser W A and Wollan E O 1951 *Phys. Rev.* **83** 333
Shull C G, Wollan E O and Koehler W C 1951 *Phys. Rev.* **84** 912
Sturm W J 1947 *Phys. Rev.* **71** 757
Weinstock R 1944 *Phys. Rev.* **65** 1
Zinn W H 1947 *Phys. Rev.* **71** 752

2   The First Diffraction Experiments: The Three Papers of 1936

2.1   The Diffraction of Slow Neutrons by Crystalline Substances*

W M Elsasser

WE HAVE examined, from the theoretical point of view, the effect on the elastic scattering of neutrons of interference effects caused by the regular arrangement of atoms in crystals. The wavelength of a neutron which possesses energy kT appropriate to thermal motion is about $1 \cdot 8 \times 10^{-8}$ cm and it is therefore quite close to the interplanar constants of many simple substances.

The selective reflections of the waves in a crystal are given by the well-known Laue–Bragg conditions

$$\kappa \sin \theta = \pi \left| l_1 \vec{b}_1 + l_2 \vec{b}_2 + l_3 \vec{b}_3 \right| = \pi B.$$

Here κ is the wave number, θ the Bragg angle—which is a half of the angle through which the neutron is scattered: the vectors $\vec{b}_i$ are the axes of the reciprocal lattice: l_1, etc., are whole numbers. We shall assume that we are dealing with a polycrystalline powder on which falls a parallel beam of neutrons of different velocities. A single nucleus produces isotropic scattering: the number of neutrons scattered by the powder within an element

---

*Translated from *Comptes Rendus, Academy of Sciences of Paris*, **202**, 1029–30 (1936).
[1]Laue, M., *Zeits. f. Kristallogr.* **64**, 115 (1926).

$d\omega$ of solid angle is given by[1]

$$I = \sum_l \frac{\pi^2 |S_l|^2 M[\bar\sigma]}{V \kappa_l^2 \sin^2\theta} f_{(\kappa, \, l)} \frac{d\omega}{4\pi}.$$

The sum includes all the orders l of reflections which make a contribution for a certain angle θ, S_l being the structure factor of each reflection. M is the number of unit cells which comprise the scattering sample, V is the volume of a unit cell, $[\bar\sigma]$ is the sum of the scattering cross-sections contained in a cell and, when several isotopes of the same element exist, it is the average sum: finally, $f(\kappa)\,d\kappa$ is the number of neutrons included in the band of wavelengths represented by the interval $d\kappa$.

In order to make an approximate calculation, we have replaced the sum by an integral over the space B (the reciprocal space of the lattice) an integral which extends over the space outside a sphere of radius B_0. Let us assume now that $f(\kappa)$ takes the form of a Maxwell distribution, which is a close approximation to the experimental facts[2]. The formula which follows gives the ratio of the number of scattered neutrons $\mathscr{I}_I$, to the incident number $\mathscr{I}_I{}^0$. Usually one merely measures the activity induced in a rather thin surface layer of a detecting substance and in this case the probability of detection of a neutron will be inversely proportional to its velocity. We have made an effective calculation by introducing first of all a factor $1/\kappa$ in the expression for the scattered intensity, which will be finally denoted by $\mathscr{I}_{II}$. We find as a result of simple calculations

$$\frac{\mathscr{I}_I}{\mathscr{I}_I{}^0} = 8\sqrt{(\pi)}N\sigma\left[\int_{B_0\sqrt{\alpha}}^{\infty} e^{-y^2}dy + B_0\sqrt{(\alpha)}e^{\alpha B_0^2}\right]\frac{d\omega}{4\pi}$$

$$\frac{\mathscr{I}_{II}}{\mathscr{I}_{II}{}^0} = N\bar\sigma e^{-\alpha B_0^2}d\omega,$$

where
$$\alpha = \frac{h^2}{8\sin^2\theta\, mkT}$$

m being the mass of the neutron, N the number of scattering centres present and $\bar\sigma$ the mean cross-section of a centre. For B_0 we can, at least in the case of simple cubic lattices, take the value corresponding to the lowest selective reflection which occurs. We shall define a critical angle by the relation $\alpha B_0^2 = 1$. This

[2]Dunning, J. R. *et al.*, *Phys. Rev.* **48**, 704 (1936).

gives $\theta = 26°$ for iron, $\theta = 26°$ for nickel, and $\theta = 25°$ for copper, whereas for the greater part of the other metals the interplanar spacings are larger and the critical angles smaller. For angles smaller than these, then $\mathscr{I}_{II}$ rapidly becomes very small. The experiments begun by Halban and Preiswerk seem indeed to indicate the existence of such an effect.

2.2 Experimental Proof of Neutron Diffraction

H von Halban Jnr and P Preiswerk

reported at a meeting of the Academy of Science of Paris on 6th July 1936 and published in *Comptes Rendus Acad.-Sci. Paris* **203**, 73 (1936).

In considering, in terms of Louis de Broglie's theory, the wave associated with a corpuscular neutron, Elsasser (1936) has suggested that a pattern analogous to that of Debye and Scherrer could be obtained even with un-monochromatized neutrons.

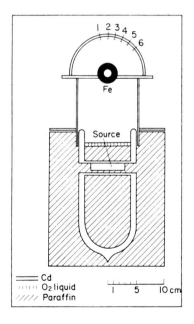

FIG. I.

(a) The distribution of neutron velocities from a source of slow neutrons (radium and beryllium, surrounded by paraffin) shows a maximum for neutrons of thermal energy, with an approximately Maxwellian distribution.

(b) The absorption of thermal neutrons in the nuclei of certain detecting elements takes place with a cross-section which is inversely proportional to the neutron velocity.

Elsasser has calculated the angular distribution of the neutrons from a source which satisfies condition (a) and which are scattered by a polycrystalline powder, assuming that they are diffracted because of their wave nature. The calculations show in particular that, for a powder of iron, the coherent neutron scattering, which can be observed using a detector which satisfies condition (b), is negligible within a cone of semi-vertical angle 26° about the direction of incidence. This angle depends on the average velocity of the neutrons and varies from a value of 18° when the moderating paraffin has a temperature of 300°K to 33° for a temperature of 90°K. The experiment consists of observing the angular distribution of the neutrons for these two temperatures.

Figure I shows the experimental arrangement. A 1 curie radium-beryllium source is placed in a Dewar flask which is filled and surrounded with paraffin. Two cadmium slits produce a collimated beam of neutrons with an angular divergence of 27°. A cylinder of iron 3 cm in height which acts as the scatterer is placed at the second slit and detectors with a surface area of 0·5–2 cm² are placed at intervals of 13° around the iron cylinder in a cadmium chamber. We have chosen dysprosium as the detecting material

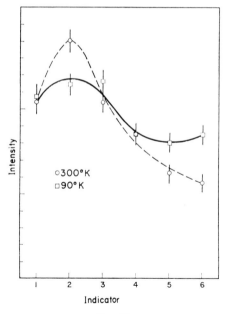

FIG. II.

because it is very sensitive to thermal neutrons. On the other hand cadmium absorbs all the neutrons which are captured by dysprosium so that there is a suitable means of completely protecting the detectors. However, the observed effect will be reduced since it can be shown by variation of tempera-ture that the cross-section of dysprosium for thermal neutrons varies less quickly than according to $1/v$.

The observed results are shown, with their probable statistical errors, in the two curves of Figure II. When no scatterer is in position, it is found that the distribution is independent of the temperature of the paraffin. The curves show that the angular distribution from the scatterer depends on the velocity of the neutrons. When the velocity of the neutrons is reduced, the scattered intensity at small angular deviations gets less, but for larger deviations it increases. The effect is a small one but, being outside statistical error, it gives clear evidence of the diffraction effect for the wave associated with the neutrons. Quantitatively we cannot compare these results with Elsasser's calculations: in our experiment the sensitivity of the detectors varies less quickly than $1/v$ and the incident beam necessarily needs a large angular divergence because of the weak neutron intensity of the source.

2.3 Bragg Reflection of Slow Neutrons*

D P Mitchell and P N Powers

THE peak of the velocity distribution of thermal neutrons[1] indicates a momentum for which the de Broglie wavelength, h/mv, is approximately 1·6Å. If such neutrons suffer Bragg reflection, they will be regularly reflected from a magnesium oxide (MgO) crystal ($2d = 4·0$Å) when the Bragg angle is about $22°(\eta\lambda = 2d \sin\theta_\eta)$.

Sixteen well-formed single crystals of MgO, about $8 \times 25 \times 44$ mm, were mounted in a ring with the source and detector placed on the axis for a grazing angle of $22°$, as shown in Fig. 1.

Background, N_{Cd}, of High Speed Neutrons

The detectors were (1), an ionization chamber filled with BF_3 in the first run, and (2), one lined with B_4C in the second and third runs. The sensitivity of both of these chambers[2] extends to neutrons of such high velocity that it was quite impossible to

*Physical Review, 50, 486–7 (1936).

[1]J. R. Dunning, G. B. Pegram, G. A. Fink, D. P. Whitehall and E. Segrè, Phys. Rev. 48, 704 (1935).

[2]Dana P. Mitchell, Phys. Rev. 49, 453 (1936).

absorb all detectable neutrons emerging in the direction of the chamber. These together with those scattered from the general surroundings account for the number (N_{Cd}) of neutrons counted when the cadmium screening of the chamber is completed by a sheet of Cd across its front.

The entire removal of the crystals made practically no change in the count of these high speed (Cd penetrating) neutrons, and further, the subsequent removal of the Cd from the front of the chamber made no appreciable increase in the count.

Neutrons Scattered by Single Crystals

It thus appears (1), that the crystals do not significantly affect the amount, N_{Cd}, of high speed neutrons counted, and (2), that the slow speed neutrons counted were scattered from the crystals.

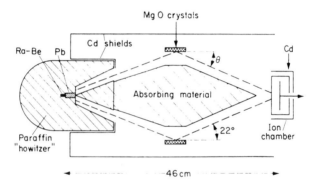

Fig. 1.

When the crystals are in the Bragg positions, the total number N_B of neutrons counted will be the background N_{Cd} plus both those regularly reflected and incoherently scattered by the crystals. The amount of incoherent scattering should be practically independent of crystal orientation, so to observe this without regular reflection the crystals were tilted, alternately clockwise and counter-clockwise, about 25° from the Bragg position. In this case of crossed crystals the total count N_X will be N_{Cd} plus the incoherent scattering. Hence $N_B - N_X$ should be a measure of the number of slow neutrons that are regularly reflected. In the first run, $N_B - N_X$ was eight times that accountable on the basis of statistical fluctuations and in the second run, six times. These results at once indicated the Bragg reflection of slow neutrons. As a check, it seemed necessary to determine by actual

TABLE I. OBSERVED NUMBERS

Run	Bragg Counts $\times 10^{-3}$	Bragg Rate N_B/min.	Crossed Counts $\times 10^{-3}$	Crossed Rate N_X/min.	Background Counts $\times 10^{-3}$	Background Rate N_{Cd}/min.
1st	23	$60\cdot5\pm\cdot4$	21	$55\cdot6\pm\cdot4$	$3\cdot8$	$43\cdot3\pm\cdot7$
		$N_B-N_X=4\cdot9\pm\cdot6$				
2nd	11	$28\cdot8\pm\cdot3$	$8\cdot6$	$26\cdot5\pm\cdot3$	3	$20\cdot9\pm\cdot4$
		$N_B-N_X=2\cdot3\pm\cdot4$				
3rd	12	$37\cdot6\pm\cdot3$	12	$37\cdot7\pm\cdot3$	6	$28\cdot0\pm\cdot4$
		$N_B-N_X=\cdot1\pm\cdot4$				

TABLE II. RELATIVE NUMBERS

Run	$\dfrac{N_B-N_X}{N_B-N_{Cd}}$	$\dfrac{N_X}{N_{Cd}}$	$\dfrac{N_B}{N_{Cd}}$
	With MgO Crystals		
1st	$0\cdot40\pm0\cdot06$	$1\cdot28\pm0\cdot02$	$1\cdot40\pm0\cdot02$
2nd	$\cdot41\pm\cdot09$	$1\cdot27\pm\cdot03$	$1\cdot38\pm\cdot03$
	With Al Blocks		
3rd	$0\cdot01\pm0\cdot04$	$1\cdot34\pm0\cdot02$	$1\cdot34\pm0\cdot02$
	corrected	$1\cdot22$	$1\cdot22$

test whether or not polycrystalline blocks of about the same size and scattering power would, due to the change in geometric disposition, scatter more slow neutrons in the "Bragg" position than in the crossed position.

Scattering by Polycrystalline Blocks

In this, the third run, aluminium metal blocks of rectangular size and thickness equal to the rectangular boundary of the somewhat irregular single crystals, used in the first two runs, were mounted in place of the crystals. Aluminium was used since it has approximately the same effective scattering power per unit volume as MgO. The result

$$N_B - N_X = 0\cdot1 \pm 0\cdot4$$

indicates that the change in geometry could not have changed the incoherent scattering much more than the statistical fluctuations.

Summary of Results

The greatest statistical precision was obtained for N_B and N_X, and hence these two values in Table I are the best evidence for the

Bragg type of reflection of neutrons. It is also interesting to note the relative amount of reflection and scattering as shown in Table II.

The correction shown takes account of the fact that the total volume of the aluminium blocks used in the third run was 1·63 times that of the crystals (due to the irregular outline of the crystals, as mentioned above).

On the basis of this evidence it seems reasonable to conclude that we have in these experiments observed the reflection of slow neutrons in accord with the Bragg relation between the de Broglie wavelength of these neutrons and the grating space of these crystals.

It should be noted that the experimental arrangement (see Fig. 1) permits a sufficiently large angular divergence so that the Bragg conditions are satisfied for a large portion of the velocity range in the Maxwellian distribution[1] of the thermal neutrons.

Grateful acknowledgment is made to Mr. Raymond Ridgeway for supplying us with the unusually large single crystals of MgO used in this work.

Pupin Physics Laboratories, Columbia University,
August 17, 1936.

3   Looking Back: Early Days and Memories

3.1 Early Neutron Diffraction Technology

C G Shull

Massachusetts Institute of Technology
Cambridge, Massachusetts, USA

The development of neutron diffraction and scattering as a powerful tool of investigation in condensed matter science became possible with the advent of chain-reacting systems operating with thermalised neutrons that induced the nuclear fission reaction. Fermi and his colleagues had first succeeded in demonstrating the feasibility of such a system with the famous Chicago Pile late in 1942 and plans were immediately formulated to build much larger systems or 'piles' as they were named at the time. The earliest of these was the Clinton Pile at Oak Ridge (or X-Pile, later known as the Oak Ridge Graphite Reactor) which became operational in late 1943 and which was to serve as the pilot plant for a series of much larger wartime fissionable-material production plants. It was a large air-cooled, graphite-moderated, natural uranium-fuelled pile that operated at a power level of 3.5 MW and produced a thermal neutron flux of about 10^{12} neutrons $cm^{-2}s^{-1}$ in its rather large volume.

It was an impressive facility, being roughly cubic in shape with edge length about 7 m and surrounded by a concrete shield of thickness 2.5 m. Cooling air was drawn through channels in the graphite moderator past lattice rows of uranium rods (2.5×10 cm) sealed in aluminium cans for oxidation protection. The rushing air, drawn by powerful fans over several hundred rows of fuel elements, generated a pronounced hissing noise throughout the building. This method of fuel cooling, long since discarded in favour of contained liquid flow, was not without hazard: occasionally leaks would develop in the aluminium sealing cans wherein the contained hot uranium metal would oxidise and rupture the can, thus blocking air flow around neighbouring cans and cascading the action. Prompt scramming of the reactor upon recognition of unusual radioactivity in the air stream was necessary to minimise the resultant damage.

Two of the four side faces of this facility functioned for service operations, with the front face being a fuel-loading area and the back face being a fuel-cannister discharge and handling area. The two transverse faces were meant for experiment access to the reactor and open, channel areas or ports were provided in the graphite moderator stacking for insertion of facilities or release of radiation beams to the outside. These channels were of standard cross-sectional area (4×4 inches) and extended the full width of the facility from face to face. It was common practice to position a 'graphite stringer' of length 4 or 5 feet and filling the port area at the centre of the facility. This would serve as a neutron scattering source to both ends of the port in a way predating modern usage of tangential ports.

There was no satisfactory way of anchoring the scattering-source plug in the channel other than through its weight in the channel and this was sometimes the cause of consternation to experimenters using neutron radiation emanating from it. The author recalls an incident that occurred in the earliest days where an unexpected drop of 30–40% in pattern intensity was encountered. After spending several days checking suspicious contributors such as monochromator alignment, counter integrity, and electronics behaviour, we found that our scattering-source plug had moved, unknown to us, to a new position in the through-port. A user at the other end of the port whom we seldom saw had made some modifications in his insert facility with the result that the central plug had repositioned itself in the cosine-law flux distribution. A graphite plug in a graphite channel with generous antisymmetric air-streaming is somewhat like an oil bubble in a U-tube!

During the war period, a single-axis spectrometer had been used at one of these ports to supply monoenergetic neutrons for transmission cross section studies. At the end of the war Ernest Wollan decided to set up a two-axis spectrometer for full diffraction pattern study and the first observations with this were made in the last months of 1945 and in early 1946. He used a rather sizeable crystal of NaCl (1×5×10 cm) as a monochromator: synthetic halide crystals were generally available at that time because of their use in infrared optics. This was mounted on a simple platform, with rotational positioning being controlled by a transverse rod that passed through a surrounding shield of stacked paraffin and lead blocks (about 3×3×2 feet). Neutrons from the monochromator were led through a channel ($\frac{3}{4} \times 1\frac{1}{2}$ inches) in the shield to an outer spectrometer which Wollan had arranged to have sent down from the University of Chicago.

This spectrometer, which was the only component of professional appearance on the assembly, had been used 8–10 years earlier by Wollan in his thesis work on gas scattering of X-rays under Arthur Compton at Chicago. It provided two coaxial axes of rotation, one for specimen support and the other for a detector with each having well cut angular scales and provision for gear coupling of the two if desired. The sleeve bearing support of the detector arm was not adequate for handling the weight of an oversize neutron detector with its local shield and was strengthened by the use of four flexible inclined cables running to a swivel-bearing in the platform ceiling directly above the spectrometer axis. The author remembers well the troubles that were encountered in getting and keeping this assembly aligned for smooth angular motion of the detector arm. It can be guessed that the base spectrometer bearing suffered under this mistreatment.

The neutron detector was a laboratory-made BF_3 gas counter (2 inches diameter, 20 inches long) and all of the electronic components, high-voltage supply, preamplifier, amplifier and scaling circuit were of vacuum tube construction made by a laboratory instrumentation group. Count accumulation was in the binary mode terminating in a mechanical register. Local shielding around the detector was initially in the form of paraffin slabs, cadmium sheeting, boron-carbide plastic layers and even some lead blocks to the extent permitted by loading on the arm.

Operation of the facility was entirely by hand and it was a time-consuming chore for Wollan and his early colleague, R B Sawyer, to collect what were the first neutron diffraction patterns of NaCl and light and heavy water in the early months of 1946 as reported in classified laboratory monthly progress reports†. The NaCl pattern was taken with a pressed briquet of powder in the form of a 2 inch diameter disc of thickness ½ inch arranged in symmetric Laue geometry and gave information on the relative scattering amplitudes of the elements and their signs. No analysis of the water patterns was attempted although the pattern differences did suggest that coherent scattering of hydrogen was being seen.

Sawyer returned to academic teaching in the spring of 1946 and the author joined Wollan in June to continue the work‡. Much of the time was spent in attempts at improving the signal intensity relative to background level through use of more effective shielding and collimation at the detector and in the shield surrounding the monochromator. Stability of counting was always a problem in the high-humidity Oak Ridge environment since air conditioning was impossible with the large air flow into the building of the air-cooled reactor. At a somewhat later time, some local

† See C G Shull *Physics with Early Neutrons*, Conference on Neutron Scattering, Gatlinburg 1976, published by Oak Ridge National Laboratory.
‡ A photograph of the author with E O Wollan at the second spectrometer, in 1950, appears as figure 3.1.

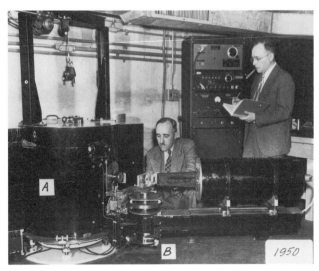

Figure 3.1. E O Wollan (left) and C G Shull with the second Oak Ridge spectrometer, 1950.

areas of experimentation on one face of the reactor were compartment-alised after sealing against leakage of the air being used for cooling the reactor and localised air conditioning was provided.

On one occasion when Wollan and the author needed to open the bulky monochromator shielding for making some modification, we were dismayed at the appearance of the NaCl monochromating crystal. Upon removing it and its support mount, we found the front Bragg-reflecting face to have a highly mottled surface, the crystal plate (originally transparently clear) had a dark, dirty-brown colour, and it was corroded firmly to its mount. Obviously the combination of air humidity, some water drippage from facilities above us, and radiation damage had taken its toll. We felt certain that our monochromatic beam must have suffered in this ordeal! To our surprise, after resurfacing the crystal, cleaning up the corrosion, heating it mildly to remove the radiation damage and recover its transparency, and remounting it, we could not claim that its neutron reflectivity had suffered at all. Anyway this encouraged us to think of other monochromator candidates which came somewhat later with use of metallic monochromators.

Along with these powder pattern studies we were interested in obtaining photographic recording of diffraction patterns and after several attempts with different white beam collimators and different photographic sensing schemes, we were able to get a clear image Laue pattern from a NaCl crystal, as shown in figure 3.2. This was obtained with a direct reactor beam (about 6 mm diameter) collimated through a heavy shield plug and

Figure 3.2. The first neutron Laue photograph, for a sodium chloride crystal, taken at Oak Ridge.

recorded on X-ray film with a neutron-absorbing indium sheet placed adjacent to the film with an exposure time of about 16 hours! There are several interesting features on this photograph aside from its being the first neutron Laue pattern: all of the diffraction spots are doubled, and it also shows the first radiographic imaging with neutrons. Not having an indium sheet of sufficient area available at the time, we had attached several strips together with Scotch tape, and there is a clear neutron imaging of this tape in the photograph. The doubling of the neutron spots puzzled us but we soon found that this was due to the existence of layers of increased mosaic spread on both faces of the crystal.

A major improvement in the mode of data collection with the first spectrometer came at the end of 1946 with the installation of the first stages of automatic operation. Credit for this must go to George Morton from RCA Laboratories who spent the year 1946–7 as an attendee of the Oak Ridge Training School which was organised to disseminate project information to a blue-ribbon group of representatives from industry and academia. Along with his School activities, Morton spent some of his time with us and, after tiring of the early hand operation of the spectrometer, he decided to design an automatic scanning system. This took the form of a motor drive on the worm-gear which turned the detector arm and associated cam wheels, microswitches and relays which permitted step scanning of the detector position. Equally important was the associated automatic printing of intensity data: the only device for doing this that we knew of

was a Trafficounter unit which as its name implies was used in that period
for monitoring street traffic flow with a pneumatic tube crossing the street.
For its intended use, this number recording unit was powered by heavy
batteries and after changing this feature it was modified to accept impulses
from our mechanical register and to print this information on command
from the scan-microswitch assembly.

In spite of the frequent servicing needed in keeping this automatic
system operational, it is hard to overemphasise the importance of its
development in permitting quantitative study of neutron diffraction patterns. The technology that went into this control system will certainly seem
trivial by present day standards but it was far from that at the time. The
author recalls the satisfaction given to us in the recognition that neutron
technology, crude as it was, had advanced in small measure beyond available X-ray technology: continuous-scanning X-ray spectrometers with electronic detectors providing counting rate recording on chart recorders were
available at the time but in principle this did not provide distortion-free
pattern study that step scanning allowed. Since background intensity was
always a problem, the neutron system was shortly modified to allow
background stages in the counting sequence. A cadmium flag on the end of
an inverted pendulum rod was moved in and out of the incident beam by
activation of a solenoid relay. After leaving the spectrometer overnight,
the author well remembers the first servicing attention being given in the
morning to the question of whether the flag position was hopefully in
proper phase to the print-out record of intensity values and angular
position!

With this automatic system, we were free to lay out a programme of
study without the prospect of tedious hand collection of data. Although
there was available at the time quantitative tabulation of thermal neutron

Figure 3.3. C G Shull at the Gatlinburg conference in 1976.

scattering cross sections, as distinct from thermal neutron absorption cross sections, the concept of coherent scattering by crystalline lattices was yet to come. The significance of isotopic and nuclear spin incoherence, of temperature disordering effects, and of nuclear recoil effects needed experimental investigation. Indeed it was not clear that a given nucleus was to be characterised with a unique scattering factor independent of crystalline environment as had been the case in the earlier X-ray diffraction studies. Even Eugene Wigner once expressed doubts about this to Wollan and the author. Serious consideration of this prospect, with its implications damaging to any future neutron diffraction technology, was needed when the earliest patterns showed inconsistencies in nuclear coherent scattering from one material to another. Gradually these puzzling features became clarified with further study and with further improvements in the technology. It goes without saying that this progress would have been greatly delayed without the availability of this first simple control system on the spectrometer.

3.2 I Remember Brookhaven

R J Weiss

King's College, University of London, UK

When I arrived at Brookhaven in November 1948 to do my thesis I was assured that the graphite reactor would be on-line in a few months. This estimate was low by an order of magnitude. During this wait my education into diffraction theory began and sharing an office with S Pasternak he kindly nurtured me through the intricacies of Zachariasen's book and its notation. Andy McReynolds introduced me to a rare subject for the time—solid state physics.

Brookhaven is the site of a former Army camp (Upton) dating back to World War I. The Army relinquished it after World War II to the Atomic Energy Commission as a 'safe' and remote location—73 miles from New York City. The old barracks were converted into laboratories and housing, and some remain even today.

Brookhaven was administered by the ten Associated Universities and became a gathering place for scientists on short visits. The major complaint was Brookhaven's remoteness and the infamous Long Island Railroad. (Legend has it that the site was chosen as equally inconvenient to all ten universities.) One of my friends visiting me and debarking at Center Moriches was certain the conductor called out 'Son of a Bitches' as the train approached the station.

The scientific atmosphere at Brookhaven was enchanting for a young student rubbing elbows with famous visitors like Bethe, Wigner, Schwinger, Fermi, Oppenheimer etc. I still vividly recall this excitement into which neither academic responsibilities nor big-city diversions intruded.

The long delay in reactor start-up gave me time to prepare my experiment. M Goldhaber suggested the use of the neutron resonance absorption in indium at $\lambda \sim 0.25$ Å as a position-sensitive monochromatic detector. An indium sheet in Laue geometry would record the powder-diffraction rings and a spark counter and camera would convert the indium beta emission into a position-sensitive visible signal. A second suggestion was to study neutron small-angle scattering. The equipment for both experiments was assembled, loaded on a station wagon and Andy McReynolds and I drove to Oak Ridge where Cliff Shull kindly made beam space available.

The indium experiment failed, mostly due to a lack of sufficient intensity at 0.25 Å. Perhaps one should revive the idea.

The small-angle scattering experiment succeeded and provided results for my thesis. Various elemental powders were immersed in CS_2 and D_2O liquids and differences in the index of refraction were measured and related to both the sign and magnitude of the scattering amplitude. These compared favourably with the more conventional Bragg scattering values obtained by Shull—except for chromium. This discrepancy later turned out to be due to the antiferromagnetism in the Cr_2O_3 Shull employed. Antiferromagnetism was still two years away from being discovered by Shull.

Having finished my thesis I wished to remain at Brookhaven to do neutron diffraction with the reactor that was certain to, someday, produce neutrons. But this was early 1950 (before the Korean war) and jobs were scarce at Brookhaven. I visited several establishments like General Electric, Bell Labs etc. with the proposal that neutron beam space would be made available to them if they employed me. But the question as to what advantages neutrons held over X-rays could not be answered convincingly before the discovery of antiferromagnetism. Fortunately Tom Johnson, Physics Chairman at Brookhaven, provided a letter of recommendation to a friend at Watertown Arsenal and I was hired under the desired relationship.

Some Brookhaven research in the early 1950s still remains in my memory like Andy McReynold's measurement of the gravitational fall of the neutron by employing a long flight path that took him beyond the walls and warmth of the reactor building. The experiment was successful but McReynolds concluded that it was the least accurate and most expensive technique employed for measuring g.

In a collaborative effort with McReynolds the critical angle for total reflection from a vanadium mirror was measured by employing nitrogen gas at various pressures to match the indices of refraction. Alas, it was later shown by conventional Bragg diffraction, by Levy at Oak Ridge, that vanadium had a negative scattering amplitude, so the total reflection we observed must have been due to a vanadium oxide coating!

Don Hughes came to Brookhaven from Chicago. He and Palevsky measured the critical angle for total internal reflection at a solid bismuth–liquid oxygen interface and they were able to determine the neutron-electron scattering amplitude. Fermi had previously done some less accurate measurements and had suggested a model for the neutron–electron interaction by postulating that the neutron was occasionally a proton–negative meson pair. About this time L Foldy showed that the moving magnetic dipole of the neutron produced a classical interaction of the right magnitude. To this date no one seems to have explained the absence of the Fermi effect.

After Shull reported the presence of antiferromagnetism in chromium powder I obtained a large single crystal and was examining the antiferromagnetic (100) reflection. It was about three times broader than the non-magnetic reflections. Purely as an idle gamble a plutonium filter was employed to eliminate the second order (200) nuclear contribution. (This was J Hastings' suggestion—he had to keep the plutonium in a safe as it was classified material.) After filtering the second-order nuclear contribution the remaining (100) magnetic peak was discovered to be a doublet and this led to the spin-density wave model (Corlis et al 1959). (A photograph of Corliss and Hastings at their spectrometer appears as figure 3.4, taken in 1958.) There was a sufficient number of materials with low enough absorption to permit neutron effects to be observed in the total cross sections. Theoretically one expected the cross sections to vary as λ^2 but in most cases the experimental results departed significantly from λ^2. This led to an awareness of the importance of extinction. I have since noticed that extinction is like dust on the floor—it eventually becomes sufficiently bothersome to do something about but only the most fastidious undertake this chore at the outset.

Beryllium and graphite thermal neutron filters became a plaything of the period. The neutrons filtered through graphite $\lambda > 6.7$ Å were diffracted by mica ($2d \sim 20$ Å) and the scattering by irradiation-induced defects was observed in graphite (Antal et al 1955).

Walter Knight occupied the apartment next to mine and I still remember him returning from his lab one evening to report that he picked up the

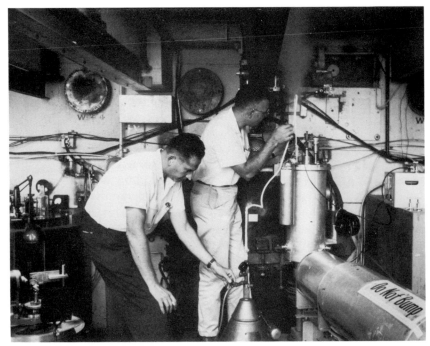

Figure 3.4. L M Corliss and J M Hastings at the Brookhaven reactor in 1958 (Copyright Gordon Park, *Life Magazine*, Time Inc. Reproduced by permission.)

nuclear magnetic resonance of the copper in his magnet coils and this was shifted relative to that of the copper in his oxide sample. This was soon followed by a special one-day conference attended by such personages as Wigner, Purcell etc to provide their concerted stamp of approval on the Knight shift.

At one point in the early 1950s John Slater and four of his students Kleiner, Koster, Parmenter and Schweinler came down from MIT for a year (Slater had to establish a NY residence to obtain a divorce). As a result of this visit we became infused with the physics of electron distributions. Slater was an eloquent speaker and my interaction with him aroused my subsequent interest in electron position and momentum distributions (and in obtaining my own divorce). On leaving Brookhaven for permanent residence at Watertown I turned to X-ray diffraction measurements of charge density, reversing the trend that most neutron diffraction scientists came from X-ray diffraction backgrounds.

Other events at Brookhaven associated with atomic energy remain in my mind. Sam Goudsmit was in the physics department and I developed a keen interest in his wartime Alsos mission, i.e. the scientific team he led into Europe after D-day to determine the extent of German atomic energy development. There emerged a public dispute between Goudsmit and Heisenberg over whether the Germans were trying to develop an atomic weapon—a position vigorously denied by Heisenberg. The principal German scientists were rounded up in early 1945 and interned at Farm Hall, an unoccupied mansion near Cambridge, England, where their conversations were secretly and (contrary to the Geneva convention) illegally recorded until after Hiroshima.

When I started playwriting some years ago I wrote a stage drama about the Alsos story and showed the play to relatives of the interned scientists as well as to one of the internees. It was in this diversion from my scientific life that I became aware that scientists make poor historians. They tend to focus in their hindsight on their own scientific achievements and less on matters of interest of historians (who are after the bigger picture, i.e. how do the *interactions* of people and ideas turn into history). Goudsmit and Heisenberg buried their differences about 1951 and I was present at Brookhaven when this occurred.

On one other occasion, just after J R Oppenheimer had his security clearance lifted by Eisenhower in that famous case, he came to an evening social event at Brookhaven as a guest of the Director. I still remember the security officers panicking at his presence. The Oppenheimer story became one of the traumatic episodes of the McCarthy era. It was perhaps unfortunate that basic unclassified neutron research had to be performed within the confines of a classified area as was the case at Brookhaven. Eventually the reactor area was totally declassified but that was many years later.

My exposure to neutron research at Brookhaven led me to remark to my boss at Watertown that a swimming pool reactor could be constructed rather cheaply. This led to a formal proposal by the Army to construct a research reactor at Watertown. Shortly after the proposal was submitted MIT announced its own plans for a reactor and I suggested we drop the Watertown effort and rent beam time at MIT. The commanding officer at Watertown suggested to the administration at MIT a joint reactor effort but MIT wisely declined US military involvement. (Figure 3.5 shows a 1964 photograph of myself with the commanding officer of Watertown Arsenal.)

The Watertown reactor was ultimately constructed and after about ten years of relative inactivity and less than a dozen research papers it was dismantled. The whole fiasco cost at least 10 million dollars. The silvery-coloured containment shell for the reactor can still be seen at Watertown—a reminder to me of that casual remark I made thirty years ago.

Neutron diffraction has revealed details about magnetism that make this property in transition metals and compounds almost intractible to theory. Slater jokingly remarked to me that neutron diffraction was an unfortunate development because he really understood magnetism back in the 1930s.

In retrospect I ask myself (for this historical account) whether neutron

Figure 3.5 R J Weiss presenting the Commanding Officer of Watertown Arsenal with a copy of his book on solid-state physics.

diffraction has paid off. Cyril Smith sat on a committee in 1946 (with Fermi) which decided that neutron diffraction was not worth support since X-rays could perform the same function. We now may laugh at this. Yet while the intricacies of magnetic structures and phonon dispersion curves have filled the scientific literature, what impact has it had on society? In such a 50 year review it is useful to take stock. I cannot answer the above question. Brookhaven will probably be around in 2036—perhaps there will be an answer at that time—if not, I will withdraw the question.

References

Antal J J, Weiss R J and Dienes G J 1955 *Phys. Rev.* **99** 1081
Corliss L M, Hastings J M and Weiss R J 1959 *Phys. Rev. Lett.* **3** 211

3.3 Early work at the Argonne

M H Mueller and G R Ringo

Argonne National Laboratory, USA

In 1943 a heavy-water moderated reactor, designed by Eugene Wigner, was built by Zinn† and his fellow workers. It was located on a 20 acre area of the Argonne Forest Preserve, south-west of Chicago and approximately 20 miles from Stagg Field, the football field of the University of Chicago and site of the first chain reaction. The reactor was called CP-3 (Chicago Pile, Mark 3) and the site was that of the Metallurgical Laboratory of the Manhattan Project. Not until 1947 was the laboratory named the Argonne National Laboratory.

Two groups led by Enrico Fermi and Donald Hughes began work on the fundamentals of neutron optics using thermal neutron beams from CP-3, though not until 1947 did the early papers of Zinn, Sturm, Fermi and Marshall reveal to physicists and chemists that neutron beams from a reactor could be used for studying solids.

Among the important early results was the absolute sign of the scattering amplitude for neutrons. This was obtained by Fermi and L Marshall (1947) who demonstrated the total internal reflection of a neutron beam incident on a graphite mirror. This result was quite important for it confirmed the application of rather straightforward non-relativistic quantum mechanics to nuclear processes, which was by no means a routine matter at the time. Fermi had been prepared for the opposite result and had actually constructed a mirror of manganese, which was known to have a sign of scattering amplitude opposite to that of most materials. Fermi, J Marshall and L Marshall (1947) built the first mechanical chopper‡ for the production of pulsed beams and time-of-flight spectra. With this a new way of producing powder diffraction patterns was demonstrated (Fermi, Sturm and Sachs 1947). These scientists were presumably more interested in nuclear physics than in condensed matter but they were very conscious of the potential of neutrons for studies of the latter and were concerned to show the similarities to, and differences from, X-ray optics. The overlap of their interests is well illustrated by the balanced index of refraction method for measuring neutron scattering amplitudes. This method was actually invented by Donald Hughes at Brookhaven National Laboratory after he left Argonne but it was first used at Argonne. The method essentially

†See photograph figure 1.2
‡See photograph figure 1.1

consists of balancing a negative and positive scattering amplitude in the composition of a mirror (in practice a liquid) until $n-1=0$ and measuring the small difference (negative) by neutron reflection. The first use of this method was for measurement of the scattering amplitude of the proton, which was of considerable nuclear interest at the time (Burgy *et al* 1951).

Hughes and his collaborators did a number of pioneering studies using the magnetic properties of neutrons (see Hughes 1954). One established that the magnetic interaction inside magnetic materials is μB, another explored the rules for polarisation by transmission, and a third studied some properties of domains in the interior of ferromagnetic materials. In the course of this work they made the first observations of small-angle scattering of neutrons. It was the work of this group that led M Hamermesh to invent the cobalt polarising mirror. (A long-delayed consequence of this early work on magnetic mirrors is the revival of the technology at Argonne by G P Felcher *et al* (1984) to study the magnetic properties of thin films using a new and more powerful analysis.)

Other work done on CP-3 which established precedents in neutron scattering includes the first study of liquids done by O Chamberlain (1950), who is famous for his part in the discovery of the antiproton. He used an ingenious diffraction geometry that made very efficient use of of the neutrons. Fortunately the great increase in neutron flux since that time has made it unnecessary to work that hard at saving neutrons. Mention should be made of the work of A H Weber and his collaborators (Brun *et al* 1957) on vitreous silica which used a novel method of analysis of neutron transmission spectra which was in some ways better than conventional powder pattern analysis. Sidhu and his collaborators started their neutron diffraction studies at the CP-3 reactor. He and McGuire (1952) showed that metal hydrides contained hydrogen (or deuterium) atoms in definite positions with respect to those of the metal atoms.

A few words about the atmosphere of the laboratory in which this work was done may be of interest. About two hundred people worked on the site which in due course came to be known as Site A of the Argonne Laboratory. It was awkward to get to, and had no nearby public transportation, was surrounded by high fences and had a serious security force. The buildings were of a style known as 'Nondescript temporary' and had been thrown together in a few weeks in 1943. It sounds about as unattractive a place as could be invented for basic scientific research but most of those working there judged it to be the best laboratory they ever knew. The intellectual atmosphere formed by scientists like Fermi, Maurice Goldhaber and Maria Mayer was extremely stimulating and the budget constraints, below what would be roughly $100 000 today, were minimal. Nobody wrote proposals to be reviewed in Washington and for the working scientists all important decisions were made on site. Even the flimsy buildings were an asset, for in an afternoon the carpenters could move a wall and enlarge a room 10 feet. Nor should the great value of possessing one of the first few research reactors in the world be neglected. The country club surroundings—the laboratory had its own tennis court—and the indifference to time clocks made for a relaxed atmosphere but nobody was deceived. Competition was strenuous and productivity was very high. Besides the great stimulus given to neutron optics the main work of the laboratory in nuclear physics was impressive. Maria Mayer worked here when she won her Nobel Prize for the shell model.

In 1954 a new research reactor went critical; known as CP-5 it was built a few miles from the original CP-3. It started with a power of 1 MW which was later increased to 5 MW. It was heavy-water cooled with an enriched uranium core, and contained twelve neutron beam ports and numerous irradiation facilities. It became a prototype research reactor for a number of installations around the world.

Sidhu and collaborators were the first group to set up their instrumentation at the new reactor. They continued their studies especially on the deuterides of Ti, Zr and Hf. In all cases only powder samples were available, hence it was essential to form the deuteride in order to minimise the background generated by the hydrogen incoherent scattering. Earlier they (Winsberg et al 1949) had shown that the phase of scattering of thermal neutrons by Ti was indeed opposite in sign to that of Zr and Hf. This led to the possibility of a nuclear null-matrix. Such a solid solution alloy consists of 62 at% Ti and 38 at% Zr. Since b_{Ti} is -0.38×10^{-12} cm and b_{Zr} is 0.62×10^{-12} cm, the resultant structure factor is zero and therefore no peaks were observed in a neutron diffraction pattern. Upon forming the deuteride the diffraction peaks appeared from the lattice of deuterium atoms only. Other light-atom studies were also carried out, for example, on UN to confirm the position of the nitrogen atoms and the rock-salt structure of this material. A number of studies were carried out by the Argonne group on non-stoichiometric metal hydrides and carbides (Sidhu et al 1963).

Crystal-structure determinations were carried out on a number of single crystals, first in the Metallurgy Division and later in the Chemistry Division. In most cases the objective was to determine the position of light atoms. Simonsen and Mueller (1965) demonstrated the power of neutron diffraction in the complete solution of a fairly complex structure ($13\times15\times7$ Å) by applying standard techniques used in X-ray determinations. Intensities from 2000 reflections were obtained by θ, 2θ scans with hand-set angles, and used for the successful solution ($R=9.8\%$) of the structure by the use of the Patterson function, which had not been used previously with neutron data. Structure determinations of a number of hydrates were successfully achieved even though nearly one-half of the atoms were H atoms (Taylor and Mueller 1965).

With increased use of neutron beams it became important to enlarge and improve the instrumentation at the CP-5 reactor. A dual unit was built with two monochromators in-line—one for a horizontal unit and the other for a vertical diffractometer. An improvement in intensity and easy choice of wavelength was possible by using a single-crystal germanium disc which had been hot-pressed to improve its reflectivity by a factor of 40 (Barrett *et al* 1963). To improve data collection for single crystals a Picker X-ray diffractometer was modified and a full circle designed and built for this instrument. Automatic angle-setting and data collection was then accomplished with an IBM card input and output system, later converted to computer control. An auxiliary piece of equipment for this dual unit consisted of a large tipping and rotating crystal mounted on a large circle (Heaton *et al* 1970). Complete single-crystal data could be obtained for a monoclinic system without remounting the crystal. During data collection on UO_2 it was held at helium temperature for six weeks.

A number of other important pieces of research were carried out at CP-5 in the 1960s. Heaton and colleagues (Gingrich and Heaton 1961) obtained scattering data on a number of liquids: data were obtained for ^7Li, Na, K, Rb and Cs, and later for vitreous silica, amorphous selenium and NaK. Rush *et al* (1966) studied the motion of water molecules and Berger (1963) promoted neutron radiography. Much research was carried out on the actinides. Studies of uranium and its compounds led to extensive work on Np, Pu, Am and Cm systems in the next two decades. Some of this is discussed by Lander and Mueller (1975).

References

Barrett C S, Mueller M H and Heaton L 1963 *Rev. Scilnstr.* **34** 847

Berger H 1963 *Non-destructive Testing* **21**, 369 (see also 1962 *Sci. Am.* **207** 107)

Brun R J, Delaney R M, Persiani P J and Weber A H 1957 *Phys. Rev.* **105** 517

Burgy M T, Ringo G R and Hughes D J 1951 *Phys. Rev.* **84** 1160

Chamberlain O 1950 *Phys. Rev.* **77** 305

Felcher G P, Kampwirth R T, Gray K E and Felici R 1984 *Phys. Rev. Lett.* **52** 1539

Fermi E, Marshall J and Marshall L 1947 *Phys. Rev.* **72** 193

Fermi E and Marshall L 1947 *Phys. Rev.* **71** 667

Fermi E, Sturm W J and Sachs R G 1947 *Phys. Rev.* **71** 589

Gingrich N S and Heaton L 1961 *J. Chem. Phys.* **34** 873

Heaton L, Mueller M H, Adam M F and Hitterman R L 1970 *J. Appl. Crystallogr.* **3** 289

Hughes D J 1954 *Neutron Optics* (New York: Addison-Wesley)

Lander G H and Mueller M H 1975 in *The Actinides: Electronic Structure and Properties* ed A J Freeman and J B Darby (New York: Academic)

Mueller M H and Knott H W 1958 *Acta Crystallogr.* **11** 751

Rush J J, Leung P S and Taylor T I 1966 *J. Chem. Phys.* **45** 1312

Sidhu S S and McGuire J C 1952 *J. Appl. Phys.* **23** 1257

Sidhu S S, Satya Murthy N S, Campos F P and Zauberis D D 1963 *Adv. Chem.* **39** 87

Simonsen S H and Mueller M H 1965 *J. Inorg. Nucl. Chem.* **27** 309

Sturm W J 1947 *Phys. Rev.* **71** 757

Taylor J C and Mueller M H 1965 *Acta Crystallogr.* **19** 536

Winsberg L, Sidhu S S and Meneghetti D 1949 *Phys. Rev.* **75** 975

Zinn W H 1947 *Phys. Rev.* **71** 752

3.4 A Childhood of Slow Neutron Spectroscopy

B N Brockhouse

McMaster University, Ontario, Canada

Personal prelude

Slow Neutron Spectroscopy or 'inelastic neutron scattering' has been a major factor in the development of the physics of condensed matter for over 30 years. This is the story of how it began, written from memory since 1982.

My education in physics commenced in 1945 at the age of 27 years. I had had a little formal education in electrical and electronic technology and some technical experience. During most of my five-year-long career as an over-age student of physics and mathematics at the University of British Columbia and of physics at the University of Toronto, it was not at all my intention to have anything to do with the neutron and neutron reactors. In my revulsion against any connection with 'atomic weaponry' I would have rejected the idea categorically. The expected usefulness of reactors for production of electrical power ameliorated the situation in my eyes so that my own rejection did not extend to an insistence that such rejection was morally binding on physicists generally. If questioned about the inconsistency of this position I would probably not have been able to justify it. But as I liked arguing with people (and I think could argue well) I would probably have come up with some sort of case. At this time I had not yet formulated the thesis: that the Laws of Physics constitute part of man's terms of reference from God, for physical action in the world, and I was still a long way from being an existentialist and thus from possessing the iron-clad defences against accusations of self/other inconsistency which this family of philosophies affords their adherents.

So it is odd that I should have ended up in 1949–50 choosing a job offer from the Chalk River Laboratories of the National Research Council of Canada over the several other opportunities open to me at the time. Perhaps not so odd if one considers the near-unique position of Chalk River in Canadian physics at the time as a place in which state-of-the-art physics was done on a large scale. Perhaps also not so odd if my own position as a married graduate student with one child is considered, a student who felt the need for more education in physics.

My thesis project and course specialisation was in the general area of low temperature and ferromagnetism. My thesis supervisors were H Grayson Smith and James Reekie, but both men left the University to assume higher positions early in my post-graduate work and I was left substantially on my own (though D S Ainslie gave me some help locally and I had correspondence with Grayson Smith). There were courses in solid state physics; these employed the famous book by Frederick Seitz and the equally famous book on metals by N F Mott and H Jones. But there were of course no references to neutron diffraction or neutron optics in the texts and few if any in the lectures.

My first exposure to neutron magnetic diffraction was, I think, in a lecture by J H Van Vleck to a meeting of the Canadian Association of Physicists, then a very new organisation. Van Vleck recounted work of E O Wollan's group at Oak Ridge National Laboratory and especially the decisive experiment of Clifford G Shull and J Samuel Smart, which verified the applicability to some paramagnetic solids of the concept of anti-ferromagnetism which had been advanced by L Néel several years earlier. From that time on I appreciated the potentialities of magnetic neutron diffraction; this undoubtedly influenced my decision to go to Chalk River.

So in late July of 1950 I travelled up to Deep River, the townsite of Chalk River, as a passenger on the truck bearing our furniture. Dorie, with baby Ann, was to stay with her parents until the small house in Deep River was shipshape. Some days later they joined me. We had intended that our stay would be a sort of post-doctorate for just a few years. As it was, we stayed for twelve years and five more children. Deep River was an almost ideal nursery and milieu for children up to teenage and Chalk River was in many ways an ideal setting for scientific research. The isolation was in each case a virtue as well as a major drawback.

The Chalk River Laboratory and the Technical Status Quo in 1950

Chalk River was a highly disciplined organisation, but one which permitted a great deal of scientific liberty. There was no difficulty for me in the matter of hours of work or in cooperation with others; I had always willingly put in long hours and tried to cooperate. But my disorganised modes of thought and work and my social clumsiness made me slow to take full advantage of the opportunities which Chalk River offered. My appointment was in the General Physics Branch of the Physics Division, to a section which was involved with neutron diffraction studies of gaseous and condensed matter and which was headed by Dr Donald G Hurst. Hurst was a Canadian, a graduate of McGill University with post-doctorate experience at the Cavendish Laboratory. He had been one of the principal architects of the NRX reactor and was to have a major place in the Canadian nuclear scene in the future. He was also the most important single influence on the development of my own career.

At the time I joined, the NRX heavy-water research reactor had then been operating for a couple of years and was reputed to have the highest flux of any reactor on earth. There was already neutron diffraction equipment operating and available for the use of our group who then comprised Hurst, Dr G H (Trudi) Goldschmidt, Dr Norman Z Alcock and myself, together with a technician, Walter Woytowich and a summer student, Myer Bloom. Bloom was working on a project to study the resonant scattering of slow neutrons by certain nuclides exhibiting high neutron absorption (Cd, Sm, Gd etc). This project I took over from him when he left in the autum to return to university and it produced our first research publications in neutron physics. Myer Bloom went on to a considerable career in Canadian physics, particularly in the fields of magnetic resonance phenomena.

To complete the vertical structure under which I served, the Head of the General Physics Branch was Dr Hugh Carmichael, an experimental cosmic-ray physicist in whose Branch I worked (until 1960 when the Neutron Physics Branch was created with me as its first head till, in 1962, I left Chalk River for McMaster University). The headship of the Physics Division had recently changed from Dr W H Watson to Dr B W Sargent, who in turn left Chalk River to be succeeded by Dr Lloyd G Elliott. Elliott's long and successful guidance of the Physics Division must be credited with a good deal of the institution's reputation in physics. Overall direction of the research programmes at Chalk River was in the hands of Dr W Bennett Lewis, perhaps the only genius I ever met, to whose initiatives the place of Chalk River in technological history must owe a great deal.

Neutron diffraction, neutron crystallography and neutron optics were already well established subjects, though not yet much exploited. The first two had developed largely through the efforts of Wollan and Shull at the Oak Ridge National Laboratory in Tennessee, neutron optics mainly through the work of Fermi and Dr Donald J Hughes at Argonne National Laboratory in Illinois. Brookhaven National Laboratory was just under development and it was 'big news' in our group when we heard that Hughes had moved to Brookhaven. The principal concepts of neutron diffraction were derived in analogy from those of X-ray diffraction, though major differences in apparatus, techniques and capability obtained and these had undergone considerable development already. Cadmium and boron carbide shielding were available and beryllium filters had been invented. Reasonable values of scattering amplitudes of many elements (and of some isotopes) for slow neutrons were in the literature, of which only a few proved later to be much in error. Several amplitudes were especially interesting; vanadium had been found to scatter almost completely incoherently and thus to show almost no Debye–Scherrer lines and light and heavy hydrogen had been shown to have very different scattering properties each from the other. These phenomena were without close analogue in X-ray diffraction. It was understood that neutron scattering amplitudes are, practically speaking in the light of nuclear theory, purely phenomenological quantities and that this will probably always be the case.

Aside from analogy, there was already available a considerable body of well established quantum-mechanical theory for neutron scattering by nuclei and lattices of nuclei, by atomic magnetic moments (in S-states) and by magnetic fields. There were still important developments to come but the major aspects of present-day neutron scattering theory were in place, though not always firmly buttressed by experiment. But computing facilities were limited to electro-mechanical calculators.

Of special importance was magnetic scattering of neutrons and magnetic neutron diffraction. Magnetic scattering amplitudes were, for the most part, already in 1950 calculable or at least guessable, in rather close analogy with X-ray amplitudes. Samuel Smart and Shull had demonstrated the feasibility of determining the magnetic structures of antiferromagnetic crystals by neutron diffraction. This was of special importance because the very existence of antiferromagnets (and of ferrimagnets) was still to some extent problematic. The possibility that I might be able to pursue studies in this area had been one of the attractive aspects of the job offer from Chalk River. As it turned out, I did do a few experiments in magnetic neutron diffraction (including the antiferromagnetic structures of Cr_2O_3 and CuO), but my main efforts soon turned in another direction.

Neutron Diffraction at Chalk River

At the time, the major thrust of Hurst's programme was the study of neutron scattering by gases. Because of the smallness of nuclei in comparison with the wavelength of slow neutrons, there is no analogue in neutron diffraction for the atomic form factor of X-ray diffraction. Thus the diffraction pattern from a collection of molecules is expected to show more marked structure with neutrons than with X-rays. This constitutes an advantage for structure studies by neutrons of polyatomic molecules. However, neutron diffraction has difficulties of its own, difficulties which originate from the recoils experienced by the nuclei from the interaction with the neutrons and from the associated Doppler shifts of the scattered neutrons. With X-ray diffraction these 'inelastic effects' are of little importance; though they exist in principle, the small energy changes involved are virtually unmeasurable in ordinary circumstances. The Compton effect is in some respects an analogue and Compton scattering must be evaluated and subtracted from X-ray patterns before analysis. But because of the small energies, there is for slow neutrons no direct analogue of the

Compton effect. Likewise, the negligible importance of the scattering of neutrons by the electrons (with their atomic form factors) comes about because of the extreme smallness of the neutron–electron interaction when compared with the neutron–nucleus interaction. All this was pretty well understood at the time.

For neutron scattering by heavy nuclei, the inelastic effects involve small enough energy transfers that analysis is feasible by methods analogous to those of X-ray diffraction. Thus, in studies of gases or liquids composed of heavy elements, the Fourier transformation method of Zernike and Prins could be employed to a reasonable degree of approximation. For poly-atomic gases 'semi-classical models' were employed in which the inelastic effects were treated as corrections in a certain order of approximation. The diffraction patterns of some simple gases had already been studied by Hurst and Alcock and quantum-mechanical calculations of semi-classical models had been made by Dr Noel K Pope of the Theoretical Physics Branch. Hurst suggested that I look into the possibility of studying the polyatomic molecule CCl_4. This project never developed because of the difficulty envisaged in containing, for weeks, the corrosive gas at high temperature and pressure, while giving access to the neutron beams. This was my first encounter with this problem but not the last. Alcock soon left Chalk River for the industrial world (and later, involvement in 'peace research') and the programme on gases lapsed. But interest in liquids continued in the group and was eventually to bear much fruit.

There were other projects underway in the group. Goldschmidt and Hurst were in the process of studying ammonium chloride (NH_4Cl) and its deuterated analogue (ND_4Cl), with a view to elucidating the nature of the second-order phase transition at $232\,K$. The transition was generally thought to involve an ordering of the ammonium tetrahedra but how was not understood nor yet what degree of quasi-free rotation was involved. At the time this was a daunting problem to be attempted with X-rays, because of the very small scattering amplitude for X-rays of the single hydrogen electron. The neutron amplitudes for both hydrogen and deuterium were known to be much more favourable and very different each from the other. It was already widely appreciated that the possibility of deuteration added new dimensions to the study of hydrogenous compounds. The crystallogra-phy of the classic problem of the ammonium halides was largely solved over the next few years, especially through the work of H A Levy and S W Peterson of Oak Ridge. The dynamics did not receive attention till many years later but is now largely understood.

As a carry-over from my thesis project, I undertook a study of the preferred orientation of the crystallites in some specimens of nickel wire which had been cold-worked and annealed in various ways. Preferred orientations in rolled aluminium foils were also studied, with a view to their use in specimen containers as windows which would not show

Debye–Scherrer lines. In many cases neutron beams are very suitable for such studies; the comparatively penetrating neutrons, in beams of large area, give better samplings of the crystallites than ordinarily do X-rays.

Finally, there was the resonant scattering project already mentioned as well as miscellaneous experiments, including some connected with development of monochromators or other apparatus.

The Idea of Slow Neutron Spectroscopy

In late 1950, Hurst arranged for our small group to meet weekly to study the theoretical literature. Alcock had left the group, but Noel Pope (a New Zealander who had been a student at Edinburgh with Max Born, the dean of crystal dynamicists) joined us. We took up for study a paper by Robert Weinstock, from the *Physical Review* of 1944, on neutron scattering by crystals and particularly by iron. We went through the paper from the beginning, page-by-page. In the course of this study we realised that the neutron inelastic scattering from a single crystal could be analysed to map out the dispersion relations of the phonons in the crystal. Furthermore the required theory would be exceptionally well founded and evocative— essentially only the conservation rules for energy and crystal momentum were involved. To do this one would have to measure the energy distributions as well as the angular distribution of the scattered neutrons. At the time this task seemed to us to be virtually impossible for reasons of intensity. We did not publish the discovery, but in Other Worlds there may be a paper giving this insight and developing the idea. We later found that the possibility had also been realised by workers at the Atomic Energy Research Establishment in Harwell, England: J M Cassels, P A Egelstaff and R D Lowde. There may have been other independent sources as well, in particular B Jacrot of the French establishment at Saclay. In the event, the proposal was first published by G Placzek and L Van Hove in the *Physical Review* of 1953, together with another proposal which we had not envisaged—determination of the frequency distribution of the phonon normal modes in a cubic monatomic crystal, by study of the intensity of the energy distribution of the incoherent scattering by the crystal. The work of Placzek and Van Hove seems to have been the inspiration for the 'cold neutron' programme of D J Hughes and H Palevsky at Brookhaven which eventually resulted in the first (beryllium) filter/chopper spectrometer.

At the time, the Chalk River NRX reactor was reputed to be the research reactor with the highest available flux of neutrons of any in the world. So I continued to think about the idea of studying energy distributions of scattering patterns and developed some notion of the considerable potentialities of such studies. I also did some calculations of expected intensities at NRX and decided that experiments were in fact feasible. Beam areas would have to be large (with consequent expecially massive shielding required) and efficiencies of all the elements of the apparatus would have to be high, but the sums came out acceptably. We then started discussions of my tentative designs for apparatus and settled on two alternatives—a triple-axis crystal spectrometer (with monochromating crystal, specimen table, analysing crystal) and an electrically phased double-chopper time-of-flight apparatus. From the spring of 1951 our thoughts focused on these two possibilities.

The triple-axis spectrometer demanded considerable improvement in the efficiencies of the two monochromators. Available data on coherent scattering amplitudes said that this ought to be achievable. The theory of extinction in crystals, including that given by G E Bacon and R D Lowde in *Acta Crystallographica* (1948), gave some quantitative guidance as to what degree of success might reasonably be expected. A programme was instituted to produce large single crystals by the Bridgeman technique, initially to involve aluminium, lead and zinc but later, it was hoped, also the still more favourable copper and nickel. In late 1952 this programme was taken over by a new arrival, Dr David G Henshaw—and by mid 1954 our monochromator problems were solved. (Henshaw was a Canadian who I had known briefly at the University of Toronto and who had taken his doctorate in solid state physics at the University of Bristol in England.) Largely through his efforts, from 1954 on we had a choice of large single-crystal ingots of aluminium with almost optimum properties or similar ingots of lead of reasonable homogeneity and mosaic spread. These could be cut on a band-saw to the dimensions required. From the beginning we had been blessed with excellent BF_3 gas proportional counters through the efforts of the members of the Counter Section, especially I L (Dick) Fowler and Harris McReady, together with Philip R Tunnicliffe. And in all our technical work we had the enthusiastic support of Walter Woytowich as well as others of the technical staff at Chalk River.

The design problems for the double-chopper spectrometer appeared formidable and detailed work awaited the accession to our group in 1952 of Dr Alec T Stewart. (Stewart was a Canadian who had taken his doctorate at Cambridge University, but had done the work for his thesis at Harwell in collaboration with another Cambridge student Gordon L Squires. Their project was the neutron transmission by ortho and para hydrogen, studied using a chopper time-of-flight instrument at the BEPO graphite reactor there. Thus Alec Stewart was the first of our group to arrive with some previous experience with neutrons.) In the event our proposed phased double chopper was never built; instead we adopted, and perhaps improved upon, the idea of Hughes and Palevsky. We built a modified 'Filter–Chopper' spectrometer in which the primary energy selection was done by matched differential polycrystalline filters of beryllium and lead. The neutron beam scattered by the specimen was then mechanically 'chopped' and the scattered neutrons analysed as to energy by their times-of-flight between the chopper and the BF_3 detector. This apparatus came into use in 1956, perhaps a year later than the beryllium filter–chopper of Carter, Hughes and Palevsky at Brookhaven. Our instrument was used for several experiments, including the phonon experiments in aluminium and vanadium. It was superseded by the first version of the 'Rotating Crystal (time-of-flight) Spectrometer' in the early autumn of 1957. (As we have seen, vanadium is a unique situation and there are good technical reasons for the choice of aluminium as the subject of a first experiment. In our case there were good practical reasons also—we had the crystals.) Phased double-chopper instruments were put into operation elsewhere however, first I think by Bernard Jacrot at Saclay. The type later received extensive development by Peter Egelstaff and others at Harwell.

But in early 1951 all this was still in the future. Almost the only experimental results in the literature relevant to slow neutron spectroscopy were transmission measurements at long neutron wavelengths which yielded total cross sections asymptotically linear with wavelength. There are normally just two important contributions to the component of the cross section which is linear with wavelength: from nuclear capture and from thermal fluctuations in the specimen. In favourable cases the measurements could be analysed to give a characteristic temperature analogous to a Debye temperature for the specimen material. Such measurements had been reported from the Columbia University cyclotron and later from 'slow chopper' measurements at Harwell, Brookhaven and elsewhere. Of this work only the series of experiments carried out by James Cassels and Gordon Squires of Cambridge University around 1950 will be mentioned explicitly, since these works were conceived in the light of the phonon scattering theory and analysed to give characteristic temperatures. This was the situation when Don Hurst and I set up to study the energy distributions of initially monoenergetic neutrons after being scattered by solids, using resonant absorption to analyse the energy distribution of the scattered neutrons.

An Infancy of Slow Neutron Spectroscopy

The experiment which Hurst and I set up utilised a modified version of our resonant scattering apparatus. Thin specimens, of polycrystalline lead, aluminium, graphite or diamond, scattered quasi-monoenergetic neutrons of 0.35 eV or so through large angles of $\sim90°$ into annularly arranged detectors. Cadmium absorbers of different thicknesses absorbed the scattered neutrons, in amounts dependent on the energy distribution with which the neutrons were gifted in the collision with the specimen. Absorption curves, the fraction transmitted of the scattered neutrons versus the physical thickness of the cadmium absorbers, held information about the energy distribution of the scattered neutrons, being ideally the Laplace transform of the energy distribution. Because the specimens were thin, multiple-event scattering was small and could be corrected for. The large range of scattering angle presented more difficulty to analysis; it was decided to compare the experimental results with model calculations, rather than to attempt inversion of the absorption curves to get experimental energy distributions. Cadmium has a large resonance at a neutron energy of 0.178 eV, some seven times thermal energy. For neutron energies from 0.2 to 0.5 eV, the total cross section shows a very steep dependence on neutron energy. Thus the absorption curves were quite sensitive to the energy distributions and the experiments for the different specimen materials gave very different results.

Two models were employed: an ideal gas composed of atoms of the appropriate mass (207 for Pb, 27 for Al and 12 for diamond and graphite) and an Einstein crystal in which the atomic masses were as above and the Einstein characteristic frequency was taken as three quarters of the Debye frequency (from specific heat measurements in the literature). The theory for the ideal gas was partly to be found in the literature; that for the Einstein crystal had been published by R J Finkelstein in the *Physical Review* of 1947. Predictably, because of the large mass, lead showed only a small, barely measurable effect, the same in both models and in the measurements. Aluminium however showed a marked effect, much the same for the two models and the experiment. Diamond and graphite showed large and rather similar effects with which agreement could be obtained only for the Einstein model by varying the Einstein frequency a bit. All this was quite reasonable, in the light of the then accepted physics of solids. For our conditions of comparatively large momentum transfer to the neutrons, scattering by lead and aluminium would usually involve production of several phonons in the specimen; quantisation might then be expected to be of little importance and the two models would be almost equivalent. For diamond or graphite, scattering would normally result in production of zero or one phonon, with little probability for multi-phonon processes.

Absorption methods were used also by workers at Harwell in two very different experiments. In the summer of 1951 a letter in *Nature* by P A Egelstaff appeared, which gave the results of experiments in which the mean energy transfers of neutrons scattered by a solid hydrocarbon were estimated from the transmissions of boron absorbers. The absorption cross section of boron is proportional to the neutron wavelength; such absorbers are much less sensitive to the energy transfers than are resonant absorbers. On the other hand, the method is applicable over a wide range of initial neutron energies. Because the specimens were thick enough to involve considerable multiple-event scattering, the experiments were qualitative rather than quantitative, but nevertheless represented the first direct observation of the energy transfers in slow neutron scattering. In 1952 a paper by R D Lowde appeared in the *Proceedings of the Physical Society of London*, which reported measurements by boron absorption spectroscopy of the neutrons involved in thermal diffuse scattering from a single crystal of iron. In the geometry employed, the diffuse streak on one side of the Bragg peak involved neutron energy gain and that on the other energy loss. These two experiments, together with that of Hurst and myself which was (published in the *Physical Review* of 1952) and one on moderating materials by G Von Dardel (published a bit later in *Archiv för Fysik*), constitute, I think, the literature of slow neutron absorption spectroscopy. More experiments using methods such as these might have been done had not the field been overtaken by events over the next couple of years. Nevertheless the work attracted some attention in the small world of neutron physics.

Some five years were to pass before I actually met Peter Egelstaff and Ray Lowde, but they had immediately assumed the status of 'our opposite numbers at Harwell' and maintained this status for many years. One of the interesting aspects of the scientific life is that it puts its members in existential contact with individuals far away in scattered places.

To continue. In 1952 we set up also a preliminary, very crude, triple-axis spectrometer. The results were inconclusive, though we did see the elastic component of incoherent scattering from paraffin and, I think, from vanadium. We still had only monochromators cut from NaCl single-crystal ingots supplied by the Harshaw Chemical Company. These we improved somewhat by various surface mistreatments, by thermal shock through immersion in liquid air and by stacking of thin slabs in parallel. But success was to await our new aluminium monochromators and the autumn of 1954.

In the meantime, Dave Henshaw had arrived to join our group, having done his graduate work in low temperature physics. With Hurst, Henshaw started to design apparatus to do neutron diffraction studies of the low temperature liquids: nitrogen, oxygen etc and their solids. The programme looked forward to similar studies of liquid helium and its mysterious modification, superfluid helium II. Thus began a research programme which, at the time of this writing, still goes on at Chalk River and which has produced, over the intervening thirty-odd years, results of the greatest importance for the physics of liquid He^4 in its two phases, I and II. From the beginning it was envisaged that eventually there would also be neutron spectroscopic studies and not just diffraction experiments, but the expected low intensities and the lack of a theoretical framework for analysis of results were viewed as formidable obstacles. These two problems were to a great extent resolved by a theoretical advance published in *Physical Review* in 1957 by M Cohen and R P Feynman. The model adopted suggested that the scattering from helium II would exhibit the Landau dispersion curve of the 'phonons' and 'rotons', in close analogy with the expectations for the phonons in crystals. Since now the intensity would be expected to fall into a 'line' spectrum, the intensity problem would be much alleviated while the analysis problem was, for the time being, eliminated. And indeed, in late 1957, experiments by Palevsky, K-E Larsson *et al* at Stockholm, by J L Yarnell at Los Alamos and by Henshaw at Chalk River, found this to be so. The helium programme at Chalk River involved successively after this time: A D B Woods, R A Cowley and E C Svensson—but this is to indeed look ahead.

Of course there were also other programmes in neutron physics at Chalk River, some of very considerable importance. But the only one to be mentioned here as relevant to these writings is that of Dr J Warwick Knowles, who pursued for many years a rather unique blend of crystal-lography, neutron physics and nuclear physics. This work used highly perfect single crystals, or pairs of such crystals, for diffraction studies of (or by means of) both γ-rays and slow neutrons. It was through the good offices of Warwick Knowles that P K Iyengar and I got access to the exceptionally large single crystals of germanium which facilitated our study of the lattice vibrations of germanium in 1957–8. Iyengar was a visiting scientist from India who joined me in this project. Our results for germanium were the first really satisfactory set of phonon dispersion curves and enabled some very pleasing correlations to be drawn with experimental results in the literature, particularly with the far infrared absorption by germanium. Of course, accuracy, resolution and sensitivity would all improve with time, but with this work (published in *Physical Review*) it could be said that the original program on which we had set out in early 1951 had reached some sort of maturity. But this too is to look ahead.

To return. By late 1952 the conceptual, experimental and theoretical foundations of slow neutron spectrometry and spectroscopy had, to a considerable extent, been laid and tested and research programmes insti-tuted. Experiments on single crystals for the purpose of determining the dispersion relation of the normal modes of the lattice vibrations (phonons) were to be carried out, the first candidate being aluminium. Experiments on polycrystalline vanadium would attempt to determine the frequency distribution of the normal modes of vanadium from its incoherent inelastic scattering. This unique material would find other applications. Elastic incoherent scattering by vanadium was to provide the zero of the scale for energy transfers and the standard for intensity calibration and for the normal resolution function. Through analogy much followed. Ferromagne-tic and antiferromagnetic magnons (quanta of spin waves) were clearly open to study in much the same way as were phonons. Indeed, demonstra-tion experiments would provide backing for the very notions of magnon and phonon, for which existing evidence was indirect. Study of fluids by slow neutron spectroscopy clearly was also potentially profitable, though methods of theoretical analysis were unclear or lacking. But analogies of liquid structure to the structure of fine polycrystals or of order/disorder in alloys, could be pursued. And near-ideal gas physics would surely find applicability. Finally, magnetic analogues to liquid and fluid could be pursued in the paramagnetic state of antiferromagnets or ferromagnets above the Néel or Curie temperature. Over the next six years the theoretical structure was considerably strengthened through the work of Placzek and Van Hove, R J Elliott, P G de Gennes, T A Kaplan, W Marshall, A W Saenz and G T Trammell—to name only a few authors of works of especial prominence. And over the same six years the greater part of the programme outlined in this paragraph had been in essence accomplished and the work published.

The experiments on aluminium and vanadium were each worked on at two laboratories: aluminium by Carter, Hughes and Palevsky, vanadium by Eisenhauer, Pelah, Hughes and Palevsky, using the Brookhaven Filter–Chopper (or Cold Neutron) facility; both metals were studied also at Chalk River by Alec Stewart and myself, using the triple-axis instrument as well as the filter–chopper method. The results of the different experiments were substantially concordant. The dispersion curves and frequency distributions obtained were crude by modern standards but left no room for doubt about the general correctness of the theory or the practicality of the experiments. With these results, published in the *Physics Review*, the *Canadian Journal of Physics* and *Reviews of Modern Physics* over the years 1955 to 1958, we have the validation of a new tool of physics and chemistry: Slow Neutron Spectrometry and the emergence of a new discipline—the study of the neutron inelastic scattering by material specimens: slow neutron spectroscopy.

3.5 Neutrons Re-cross the Atlantic

3.5.1 Reminiscences

G E Bacon, University of Sheffield, UK

1936, the year in which the neutron was first diffracted, was the year in which I first entered the Cavendish Laboratory as an undergraduate. In the following year Lord Rutherford died and was succeeded as Cavendish Professor by W L Bragg. By my third year Bragg, Bradley and Lipson were well established with the original Metro–Vick continuously evacuated X-ray machines and I was able to take my first powder photographs. I was destined to become an X-ray crystallographer and in June 1939, supervised by Henry Lipson, I embarked on my PhD course—an investigation of FeCr alloys. This activity was short-lived, for World War II started and, like so

many of the Cavendish staff and students, I found myself at the Air
Ministry Research Establishment working on 'R.D.F.' (radio direction
finding), diffracting a much longer wavelength.

As the end of the war came into sight, some of the staff were surrepti-
tiously acquired by Dr J D Cockcroft for work on atomic energy. I escaped
until February 1946 when I went to Chalk River for 3 months, to prepare to
study the Wigner effect in graphite by X-ray diffraction. On the way there I
called at Montreal University and met John Warren, whom I had known at
Malvern. He suggested that I should also get involved in neutron diffraction, about which I now heard for the first time.

At Chalk River I saw ZEEP and the preparations for NRX and then, in
July 1946, arrived at Harwell with the task of producing a spectrometer to
do neutron diffraction. All was very primitive and the mud, described by
Peter Egelstaff in the following article, was very prevalent: there was a
security fence guarding the main road but it was some time before it
encompassed the other three sides of the Establishment.

GLEEP (graphite low-energy experimental pile) operated in August
1947 and, with John Duckworth, I was able to get my first feeling for
neutron diffraction by carrying out some simple measurements with alkali
halides as monochromators. We gained an inkling of the interplay of
absorption and mosaic spread by showing that although (as reported) the
reflectivity of LiF could be increased by roughening the surface with
sandpaper, this did not happen with NaCl. For anything more elaborate we
awaited the completion of BEPO (British experimental pile—pronounced
Beepo by Cockcroft but called Beppo by everyone else). Meanwhile I
continued X-ray work on graphite, read the papers which were beginning
to come from the USA and considered what topics we should work on
when our neutrons were available. We were much attracted by the
possibilities of single crystals, which the Americans had set aside in favour
of more rapid advances with powder, and we turned our thoughts to the
problem of extinction. Following a most illuminating co-operation with
Ray Lowde, 1948 saw the appearance of the paper 'Secondary Extinction
and Neutron Crystallography'. About the same time Ron Dyer became my
assistant and began his thirty years of service to the neutron community.

At the end of 1949 our first powder diffractometer, then called a
spectrometer, was ready. It was designed and manufactured by John
Curran Engineering of Cardiff, a firm who had earlier assembled mobile
radar equipment for me. The spectrometer (shown in figure 3.6) was
unique in being mobile. The reason for this was that it had been hinted by
the reactor engineers that we might have to remove our apparatus at short
notice. However, this dire possibility never happened and the spectrometer remained permanently in place. It was by no means automated and
the main requirements for an experimenter were a stop-watch, two stout
arms for turning the handles and a lot of patience. Our first study was of

graphite and was the forerunner of what came to be called the X–N
technique. We were able to show that the reflection intensities for neutrons
were in accordance with calculation and that the long-acknowledged
anomalies in the X-ray pattern were due to an asymmetrical electron
distribution with a concentration on the C–C bonds.

Figure 3.6. The first Curran spectrometer, installed at BEPO in 1949.

Meanwhile Ray Lowde had set up a single-crystal spectrometer to use
white radiation and, inspired by a paper by Moorhouse, turned his
attention to magnetic inelastic scattering. We heard for the first time, of
'spin waves' but in relation to the neutron intensities which were available
this work was in advance of its time.

1951 was a year of change which ended our initial period of isolation. It
was the year of the Stockholm conference of the International Union of
Crystallography and in advance of this we had a visit at Harwell from Cliff
Shull. In a session of the conference held at Uppsala University many
crystallographers heard for the first time about ferromagnetic scattering
and the importance of secondary extinction. By chance the two neutron
papers in the session were combined (figure 3.7) with papers on ferroelectrics, an unconscious portent that during the next few years three independent groups at Brookhaven, Harwell and Oak Ridge would be using
neutrons to unravel the role of the hydrogen atoms in KH_2PO_4.

Among crystallographers generally in the United Kingdom there seemed
to be a strange reluctance to become interested in neutron diffraction. This

.ay Morning, June 29th

Neutron diffraction and ferroelectrics

Lecture room X

Chairmen: I. Waller
E. Wood

Paper No.	Authors	Title
N—1	G. E. Bacon	Neutron diffraction at Harwell: measurement with single crystals. 20 minutes
N—2	C. G. Shull	Magnetic crystallography and neutron diffraction. 20 minutes
F—1	I. Nitta, T. Watanabé, S. Seki and R. Kiriyama	Thermal transition in pentaerythritol. 10 minutes
F—2	R. Kiriyama and H. Ibamoto	Dielectric phenomena of K_2SnCl_4. H_2O and $K_2HgCl_4 \cdot H_2O$ single crystals. 5 minutes
F—3	R. Pepinsky and B. C. Frazer	X-ray studies of ferroelectric crystal transitions. 10 minutes

Figure 3.7. The neutron diffraction session at the 1951 meeting of the International Union of Crystallography, held at Uppsala University.

was perhaps due to reaction after the war to any activity which involved security fences and the Official Secrets Act. There were of course exceptions. J D Bernal and Kathleen Lonsdale were keenly interested at an early stage—I like to think that it was because they had been the examiners of my PhD thesis—but it was not until we had done a good deal of work on hydrogen bonds that interest was really aroused. One of our earliest supporters was J C Speakman, around 1955, who inspired a paper on potassium hydrogen bisphenylacetate and who remained a neutron disciple among the chemists for the rest of his life. More striking perhaps was the influence of Shull and Smart's 1949 paper on MnO which impressed the workers in magnetism. Notably there was Bob Street who, in 1952, began a long collaboration on magnetic materials and Terry Willis, whom we first interested at a Physical Society Exhibition and who forsook the General Electric Company to join us in 1953.

From my own personal point of view I recall particularly my visit to the 10th Pittsburgh Diffraction Conference in 1952, followed by my first visit to Oak Ridge—with the bonus of obtaining early single crystals of copper and lead which made much superior monochromators, because of their greater mosaic spread—and then to Argonne and Brookhaven. I remember also a discussion in a Boston café with Dick Weiss and others at which we agreed that p would be a good symbol for the magnetic scattering amplitude, rather than the original D of Halpern and Johnson. These were the years when the circle of workers in neutron diffraction was growing rapidly, although it was still sufficiently small for practically all to be personally acquainted. Another particular memory is a discussion with I V Kurchatov when he visited Harwell in April 1956. He did not mention any neutron diffraction work going on in the USSR and, indeed, I enthusiastically suggested to him that he should start some.

In retrospect the most remarkable characteristic of work in these early years was the simplicity of the equipment and the tools of the trade. For example, figure 3.8 shows the wooden collimator which was used for the BEPO powder spectrometer and figure 3.9 shows the very simple spectrometer used to study single crystals of chromium potassium alum. Likewise, figure 3.10 shows a very early monochromator, cut from a lead crystal given to us by Oak Ridge. It can usefully be contrasted with figure 8.6, a highly sophisticated double-focusing monochromator in use at the Institut Laue–Langevin more than 30 years later. There were no computers and our first single-crystal problems, with KH_2PO_4 and sodium sesquicarbonate, were solved by Fourier syntheses performed with Beevers–Lipson strips. We had no step-scanning of the spectrometer, although hand-turning and the stop watch had been superseded by slow continuous rotation, using a pen recorder connected to a ratemeter. The pens were very temperamental and the whole system was very much at the mercy of electrical interference from electric drills and overhead cranes operating in the vicinity. Nevertheless there were advantages: for the operator who was in possession, there were no grant-application forms to be filled in and no necessity to wait, say, twelve months for two or three days of reactor time.

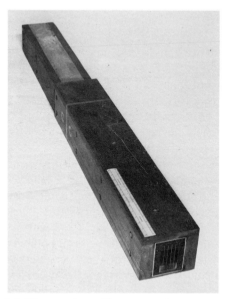

Figure 3.8. The wooden collimator used in BEPO, 1949.

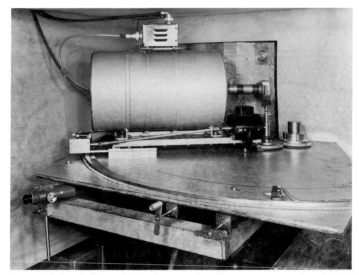

Figue 3.9 A very simple single-crystal diffractometer used at BEPO, 1955.

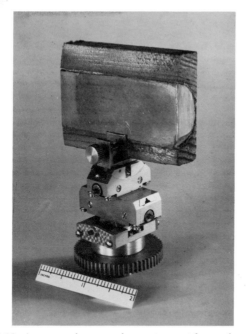

Figure 3.10. A very early monochromator, cut from a lead crystal.

Progress seems to have been fast, in spite of what would now be regarded as laughably small neutron intensities, but this belief may be

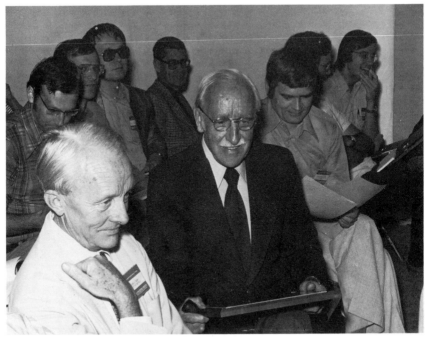

Figure 3.11. G E Bacon with E O Wollan at the Gatlinburg conference, June 1976.

deceptive. No doubt, many of the physical and chemical studies which we made were relatively simple. When many virgin fields are available for picking it is possible to make important progress with a minimum of data.

3.5.2 Early days at Harwell

P A Egelstaff, University of Guelphi Ontario, Canada

Neutrons re-crossed the Atlantic when Sir John Cockcroft left Chalk River to take up his appointment as Director of Harwell. During the early days of Harwell there was great emphasis on the design and building of the new reactors GLEEP and BEPO, and on reactor physics experiments. In those days Otto Frisch and then Robert Cockburn ran the 'Nuclear Physics' Division which looked after nuclear and reactor physics. Both they and Sir John were strongly interested in the application of slow neutrons to solid state physics, a subject in which rapid progress was being made at that time in the US laboratories. In the General Physics Division George Bacon and Ray Lowde were asked to build up neutron diffraction and were designing new instruments from scratch. John Duckworth was asked to make neutron scattering from matter an important part of his new slow neutron group, although its main emphasis was neutron cross sections for reactor design. Duckworth decided to build both a crystal spectrometer and a Fermi chopper for the cross section measurements, and two new scientists (Alec Merrison and Peter Egelstaff) who joined him were given the responsibility for each one respectively. Experimental progress was slow at first, as the laboratory was being built from the ground upwards and work on the reactors had the first priority. However some work on the crystal spectrometer was made at GLEEP in Hangar 8 during 1947, while the Fermi chopper was designed for the new reactor BEPO in Hangar 10 for late 1948. At that time the first item issued to new recruits walking between Hangars 8 and 10 was a pair of gumboots. There was mud everywhere. This seemed to come from an array of trenches which were being built in every direction. They were to become a system of ducts for heating and power, but during the first two years their mud dominated life at the lab.

Figure 3.12. P A Egelstaff testing an early chopper at Chalk River in 1957.

Improvisation was the greatest asset in doing the experiments, and electronics the biggest headache (see figure 3.12). Neutron detectors were notable by their absence. Russel Aves ran a group which built the detectors (BF_3); he made them lovingly by hand and when one was finished it was protected and cared for like a bar of gold. Amplifiers, discriminators and power supplies were made locally to begin with, and hence when one finally could detect neutrons it was felt to be an achievement. A monochromator was an asset—crystals from museums were considered highly. All data were recorded by pencil and paper. Since the equipment was not too reliable, many measurements were repeated. Data reduction was done by hand using a slide-rule, although there were a few mechanical calculators (for addition and subtraction) available. One had to apply in advance to obtain one of these for a few hours.

Duckworth's group were developing contacts with the slow neutron group (Havens and Rainwater) working on the Columbia cyclotron pulsed neutron source, when one day Sir John Cockcroft walked into Hangar 8 and said 'Why don't you build a pulsed neutron source based on the electron linac?' This remark changed the direction of our activities and led to the installation of the 3 MeV electron linac in Hangar 8 as a pulsed neutron source for neutron cross section measurements and for solid state physics. By this time Egon Bretscher had taken over direction of the Nuclear Physics Division.

Both Cockcroft and Bretscher were interested in and supportive of collaborative ventures with the universities, and they backed joint activities from the earliest days of Harwell for the remainder of their lives. The large university neutron beam programme in the UK stems from their work. At first one of the best routes was through the introduction of the Harwell University Research Fellowships. Both Jimmy Cassels and Gordon Squires of Cambridge were early holders of these fellowships. Cassels stressed the importance of using slow neutrons for studying dispersion curves of simple crystals and designed a three-axis crystal spectrometer for BEPO in 1950. Unfortunately BEPO was the worst reactor ever used for thermal scattering experiments (due to the low source flux, the extremely long and narrow holes and the fact that the beam holes passed through the thermal column which had a very thin shield, the neutron beam fluxes were lower than elsewhere and the background was higher). For this reason Cassels gave up his programme on lattice vibrations and looked to other areas of physics. Squires followed Cassels and, perhaps because of the new reactor designs being talked about in early 1950s (e.g. HIPPO, which was an improvement on NRX), decided to make pilot experiments on the available sources (electron linac and BEPO) in preparation for the future. His programme led to many Cambridge students being trained in neutron scattering at Harwell, and they have now spread through the leading neutron scattering centres world-wide. The support of Squires' group may be one long lasting result of the HIPPO design, which was dropped after the NRX accident in 1952 and replaced eventually by DIDO and PLUTO (although unfortunately they had less satisfactory designs for neutron beam work).

The low performance of British reactors led to the development of cold neutron sources. In 1952–3 Heinz London and Peter Egelstaff at the instigation of Sir John Cockcroft began a series of low-angle neutron diffraction and inelastic scattering experiments on liquid helium using 4 Å neutrons. One day London said 'this experiment is going too slowly, why can't you get more neutrons?'. 'In principle it's easy' replied Egelstaff 'sometimes at cyclotrons they use liquid hydrogen as a moderator, and all we have to do is to maintain a few 100 ccs of liquid hydrogen at the centre of the reactor. The moderation is fast in hydrogen and we should be able to lower the neutron temperature with a small quantity, but it must be hydrogen itself not a hydrogenous compound.' 'Very well' said Heinz London 'let's do it, I'll build the low temperature side and you look after the rest.' The reactor manager at BEPO was Bob Jackson, who on being approached about the safety of liquid hydrogen in BEPO said to Egelstaff 'you are the designer, engineer and physicist, you are also the safety committee, therefore your head will roll if anything goes wrong'. Careful tests and exploratory experiments were done by two new recuits Ian Butterworth and John Webb and the first cold neutron source was installed in BEPO in 1954. It could be operated so safely and successfully that later its operation was passed to the PhD students in Squires' group who used it for several years. This cold source was used also by Bill Mitchell's group of Reading University in the mid-1970s, in order to study low angle diffraction effects from amorphous and disordered solids. The advantages of high resolution small angle neutron scattering experiments were discussed and designs for DIDO and PLUTO were proposed but not approved. At this time the technology of detectors and electronics was too primitive to support the sophisticated instruments required for this field.

Even in these early days the low flux and intermittent operation of the electron linac made it less satisfactory than BEPO for solid state physics, although the pilot experiments could indicate a brighter future. The bread and butter work of Duckworth's group was neutron cross section data for reactor design purposes. After he left in the early 1960s this was actively continued on the old electron linac and later on the new 10 MeV linac, and also many measurements were made on BEPO using slow and fast choppers. The great activity in pulsed neutron work for cross section measurements, led to the idea that it could be used for neutron diffraction work. In 1954 Egelstaff gave a review of Harwell ideas to that time, for the International Congress on Crystallography in Paris. Among the audience was Dick Weiss who explained that similar ideas had been discussed at BNL. Little was done however other than demonstration experiments, until Bronislaw Buras took up the field seriously in the early 1960s.

Possibly the message of the period 1946–56 is that neutron scattering work, both diffraction and inelastic scattering, became well established in Britain at Harwell and in several universities. Many ideas were discussed for new fields of study, for new instruments and techniques. The authorities gave this field their support and blessing, but the source flux, the reactor design and the technology of detectors and electronics were all too primitive to allow for many of the developments that the enthusiasts of the time hoped for.

3.6 Recollections of a Research Student 1948–51

G L Squires

Cavendish Laboratory, Cambridge, UK

I became a research student in the Cavendish Laboratory in the summer of 1948. Although nuclear physics was not the pioneering subject it had been in Rutherford's day it was still a major interest in the Laboratory, which possessed two Cockcroft–Walton machines giving energies of 1 and 2 MeV, and a cyclotron with a 37 inch magnet capable of accelerating deuterons to about 9 MeV. I toured the groups to see what they had to offer.

The cyclotron group was run by Albert Kempton and was pursuing two quite separate lines of research; one, in which Kempton himself was mainly interested, was the study of nuclear reactions initiated by charged particles at low energy. The other, an investigation of the scattering of thermal neutrons by metals, was conducted by James Cassels, a research student two years senior to myself. It would be gratifying to say that I was inspired by the subject of thermal neutron scattering and saw its vast possibilities. The truth is somewhat different. The cyclotron group possessed an electronic neutron velocity selector, an elaborate apparatus containing about 300 thermionic valves that generated so much heat that it had to be cooled by a large fan blower. During the Second World War I had attended a crash course on electronics at Cambridge and had developed an interest in pulse generators, ring circuits and other electronic devices. Here was a glamorous electronic instrument—I made my choice.

The velocity selector had been designed by Kathirkamathamby Kandiah, and built without a circuit diagram—or so the story went. At any rate no diagrams were available and, as Kandiah had left the Cavendish

Laboratory for the Atomic Energy Research Establishment at Harwell, it was difficult to make modifications to the circuits without them. Accordingly my first task as a research student was to trace all the circuits, wire by wire, and make a set of diagrams. Some idea of the physical scale of the apparatus may be obtained from the fact that there were six units, each one being so long that when I needed to get it out of its rack to trace the wiring it took two of us to lift it.

Neutrons were generated in the cyclotron by the $^9Be(dn)^{10}B$ reaction. The way the velocity selector worked was that it produced a succession of 36 (or other multiple of 12) top-hat pulses in a cycle, which was then repeated. Each pulse was $270 \, \mu s$ long. (Other values from 10 to $810 \, \mu s$ were available.) The first of the 36 pulses was applied to the radio-frequency oscillator of the cyclotron, which was effectively on only during that pulse. The neutrons produced were moderated by a disc of wax about 20 cm in diameter and 5 cm thick placed just outside the cyclotron, near the beryllium target. About 3 m away was our single neutron detector, a splendidly dignified BF_3 chamber in a brass shield, with its head-amplifier attached. How carefully we handled it. There was no spare. If anything happened to it, our whole programme would stop dead—fortunately nothing did. The output from the detector was gated by 10 successive top-hat pulses from the velocity selector, thereby giving us 10 time channels.

The output from each channel was sufficiently low that, after going through a pair of binary counters with neon indicators to provide a scale-of-4, it could be counted by a mechanical post-office register. So the velocity selector plus the cyclotron provided a pulsed source with time-of-flight analysis, similar in principle to the machines of today—the wheel has come full circle.

The neutron source was too weak for us to detect scattered neutrons. We could only measure transmission ratios, and hence total cross sections. So all our samples were mounted on a little trolley, which was pulled in and out of the beam—by hand of course. All the neutron measurements in my three years as a research student consisted of columns of numbers, the readings from the post-office register and the two neons for each of the 10 channels, with the sample in the IN or the OUT position.

Kempton was my official supervisor and gave me much helpful advice, but during my first year I had in effect a second supervisor in Cassels. From him I learnt the good habit of not letting measurements accumulate, but of working out the results as the experiment is done. Being close in age he had no inhibitions about working me hard. I remember once, at the end of an exhausting day when we had made measurements non-stop from 10 a.m. till 6 p.m., his putting on his jacket to leave the laboratory and saying 'Make sure you work out today's results this evening, so we know what to do when we start tomorrow—I'm going to the cinema.'

I did two separate experiments during my three years as a research student. The first was to measure the total cross sections of magnesium and nickel as a function of temperature for neutron wavelengths in the range 5 to 10 Å. This was a continuation of work started by Cassels and Robert Latham, who had made similar measurements on iron and aluminium. The impetus for the work came from a theoretical paper by Weinstock in 1944, who calculated the total cross section for coherent one-phonon scattering as a function of crystal temperature and neutron wavelength. He gave a rough argument for showing that higher-phonon processes were negligible. By working at wavelengths beyond the Bragg cut-off, and hence excluding elastic processes, we hoped to measure only the one-phonon scattering and compare the measured values with those of Weinstock. Cassels found some discrepancies in the case of iron which he correctly attributed to magnetic scattering, not included in Weinstock's calculations. Aluminium had given results in reasonable agreement with the theory.

It soon became clear that the scattering in magnesium was greater than Weinstock predicted, and we guessed this was due to multiphonon processes, which, owing to the comparative lightness of the magnesium atom, were of greater significance than Weinstock had suggested. I devised a method of estimating the magnitude of the higher-order processes which, when added to the single-phonon cross section, gave satisfactory agreement with the measurements.

I would like to recall one incident in connection with the phonon work. One day in the summer of 1949 Cassels came to me in some excitement. The thought had occurred to him that the two conditions—conservation of energy and quasi-momentum—in coherent one-phonon scattering provided a method of determining the frequencies of phonons in crystals. He went to Robert Frisch, at that time the Jacksonian Professor in the Cavendish Laboratory, to ask his advice about publishing the idea. Frisch advised him to wait until he could provide an experimental demonstration of the method. Cassels, being only a research student, acquiesced. The advice, though well meant, was unfortunate. In fact it was another six years before Brockhouse and Stewart at Chalk River and Carter, Hughes and Palevsky at Brookhaven, working with improved neutron spectrometers and sources, were able to make the first measurements and thus demonstrate the method, which has become the standard one for the determination of phonon frequencies.

The second experiment on which I worked was the measurement of the total cross sections of ortho- and parahydrogen, to determine b_t and b_s, the bound triplet and singlet scattering lengths of the proton. It was done in collaboration with Alec Stewart, who had come to the Cavendish Laboratory as a research student from Dalhousie University. Although the cross sections for ortho-and parahydrogen give b_t and b_s separately, the main interest in the experiment was the determination of the coherent scattering length of the proton, i.e. the combination

$$(3b_t + b_s)/4.$$

This is given (apart from an ambiguity of sign) by the parahydrogen cross section. Previous to our experiment a large team in the United States had measured the parahydrogen cross section and had obtained the value $f = -3.95 \pm 0.12$ fm, which agreed with a value obtained from measurements on sodium hydride, but was inconsistent with the value of Hughes *et al* obtained by neutron reflection from a liquid hydrocarbon mirror.

The American ortho–parahydrogen team had had nine members, so it was somewhat presumptuous for two research students to try to repeat the experiment, but fools rush in.... Stewart and I realised that some complicated glasswork would need to be constructed to handle the hydrogen gas, and in particular to measure the ortho/para ratio. We obtained a supply of glass, and each sat down with a blow lamp. Apart from a little glassblowing as undergraduates (obligatory in those days) neither of us had had much experience, but we set to with a will. After about ten minutes it became apparent that Stewart had some talent in the matter and I had none. Fortunately there was much else to be done in the experiment, so I was happily able to leave the glassblowing side to him. In fact the glasswork became increasingly complicated, and we had to call in the services of the Cavendish Laboratory glassblower, but the fact that Stewart was able to do so much himself meant that we always got prompt and sympathetic service from the glassblower.

The measurements had to be made at liquid-hydrogen temperature (to minimise the inelastic scattering due to changes in the rotational state of the hydrogen molecule), and we were concerned at the safety aspects of hydrogen leaking in the cyclotron area with its occasional 10 inch sparks. One of my tasks was to construct a hydrogen detector, based on a heated platinum filament whose resistance changed in the presence of hydrogen. (A commercial detector was available, but it cost £50, which was considered too expensive.) My device rang a bell when the hydrogen concentration in the air reached a certain value, and usually went off when liquid hydrogen was being transferred to the scattering chamber containing the gaseous hydrogen sample. If it did not go off on these occasions, I was not above turning up the sensitivity control until it did so, thereby maintaining the confidence of the assistants that Stewart and I knew what we were doing.

This was rather necessary as we had a certain number of mishaps. On one occasion we were next to the cyclotron when someone switched on the current through the magnet coils. Stewart was holding a large screwdriver which flew from his grasp to the magnet, narrowly missing my head. On another occasion we tried to abort the experiment after Frank Sadler, the chief assistant at the Mond Laboratory which supplied the liquid hydrogen, had gone home. We tried to remove the liquid hydrogen by bubbling gaseous hydrogen through it. Something went wrong and the resulting pressure blew out the glassware—fortunately it went to the side opposite where we were standing.

Our most spectacular accident arose from my own experiment on magnesium, which I was concluding when Stewart first arrived at the Cavendish Laboratory. I wanted to make measurements with the sample at about 180 K, and found that the melting point of toluene was suitable. Accordingly I made a mixture of toluene and liquid oxygen. (Liquid nitrogen was also available, but we were encouraged to use the former as it was cheaper.) Stewart was worried, rightly as it transpired, about the safety of my procedure. He therefore made a mixture, with very small quantities, of the ingredients I was using and set a match to it. There was a loud explosion. He was severely shaken, and I had to take him to the Casualty Department at Addenbrooke's Hospital. Again, fortunately, it turned out that he had suffered no serious damage, other than to a luxuriant, chestnut moustache with which he had arrived at Cambridge. It was so severely singed that he shaved it off and remained moustache-less for the rest of his stay. I absorbed the lesson and used liquid nitrogen thereafter.

One feature of the hydrogen experiment was that the Mond Laboratory could only provide the liquid hydrogen once a week. Once it was in the scattering chamber and the sample had cooled to 20 K, we made the neutron measurements for as long as the liquid hydrogen lasted—usually about 18 hours. In this time we carried out about 20 runs of alternate sample IN and sample OUT measurements, 30 minutes for the former and 20 minutes for the latter, plus a few background runs with a cadmium sheet blocking the beam. Apart from recording the numbers in the 10 time channels we were continuously monitoring the ortho/para ratio by a thermal conductivity method. The liquid hydrogen usually ran out at about 12 noon after our all night session, and we trundled home feeling tired and virtuous. Our final value for the coherent scattering amplitude for the proton was -3.80 ± 0.05 fm, in good agreement with the liquid-mirror value at that time, and just consistent with the present accepted value of -3.7423 ± 0.0012 fm.

Looking back at those times I have two main impressions of how things have changed. Firstly, we had the equipment in our own laboratory, and we built our own apparatus. It took time, and no doubt we did it inefficiently. But we learnt a lot, especially from our mistakes. We knew the apparatus thoroughly, and if anything went wrong, we were to blame and suffered the consequences. But as time went on we made fewer mistakes, the results assumed a consistency, and our confidence increased. Secondly, we did nearly all the calculations with a slide-rule. When I became a research student I celebrated my improved status by purchasing an Aristo Multilog slide-rule. It was a de luxe model with several exponential and folding scales, and was priced accordingly—£3.15.0.

I do not hanker after the old conditions; with present-day sophistication of techniques, shared instruments, run by committees, are inevitable. One must use computers to design the apparatus, control it when built, and work out the results ('process the data' in modern jargon). But even today I believe it is good practice to do some of the calculations by hand, to get the 'feel' of the results—before bringing in the computer. But whatever the pros and cons of the different ways of working, my research student days were a productive and enjoyable period in my life, and, I think, in those of my colleagues at the time.

4 World-wide Spread of Neutron Scattering: A Miscellany of Stories

4.1 Neutrons in The Netherlands and Scandinavia

A F Andresen† and J A Goedkoop‡

†*Institute for Energy Technology, Kjeller, Norway*
‡*Netherlands Energy Research Foundation, Petten, The Netherlands*

Norway and The Netherlands

In the early days of 1947 the Dutch theoretical physicist H A Kramers travelled to Scandinavia to explore possibilities for cooperation in nuclear science, in particular to find an application for about eight tons of uranium concentrate purchased by the Netherlands government just before the German invasion and still hidden from the public eye. At Kjeller, just outside Oslo, he found a group led by the astrophysicist Gunnar Randers and one-time polar explorer Odd Dahl building a nuclear reactor. They were to obtain seven tons of heavy water for slowing down the neutrons from the Norwegian firm Norsk Hydro, Rjukan. By virtue of the very small neutron absorption in that material, they would need only about 5 tons of natural uranium. However, the extraction of this quantity from a mine in southern Norway was lagging far behind schedule.

That same day Kramers and Randers drew up the principles of an agreement under which the Norwegian reactor would be fuelled with the Dutch uranium and become the focus of a joint research institute. Subsequently, a deal was made with the United Kingdom Atomic Energy Authority involving an exchange of the Dutch yellowcake for fabricated metal rods, allowing the reactor to be rated at 100 kW, much higher than would have been possible if, as had been planned, uranium oxide had been used. The Dutch–Norwegian cooperation was formalised in 1951 through the formation of the Joint Establishment for Nuclear Energy Research, JENER.

JENER. The reactor, JEEP (Joint Establishment's experimental pile) went critical the same year and was then the first reactor in a country outside the big powers. In those days information on reactors and activities around them was highly classified, and it was stressed that the research at Kjeller and its results should be open to everyone. As a result a truly international cooperation developed. Soon after the start a strong group of Dutch scientists began to arrive. One of them, J A Goedkoop, a student of C H MacGillavry, professor of chemical crystallography in Amsterdam, had applied specifically to do neutron diffraction. His first experiments with neutrons, carried out in the autumn of 1952, were to produce some Laue photographs showing to crystallographers in both countries that indeed there were neutrons available, and of sufficient intensity to do crystallographic studies. Goedkoop had expected interest from the strong crystallographic group of Odd Hassel at the University of Oslo, but now found that relations between that group and Kjeller were somewhat strained. Being an outsider was thus an advantage, and with the cooperation of N Norman of the Physics Department at the University of Oslo funds were raised for putting up a general purpose diffractometer at the JEEP reactor. A student, A F Andresen, was assigned to the project.

During one of his trips to America Randers had been able to persuade A W McReynolds to join JENER for a year. McReynolds brought with him a lot of experimental know-how, including a complete set of drawings of a step-scanning goniometer he had built at the Brookhaven National Laboratory. It was now decided to use the money from the University for building a similar instrument, the most memorable feature of which was a piece of bicycle inner tube for eliminating play from the gear train. When finished it was placed at a neutron beam hole (figure 4.1) which had been equipped with one of the lead monochromator crystals that by now were grown at Delft.

The first structural problem to be solved with this diffractometer was that of copper hydride, a problem which had been suggested to Goedkoop before he left Holland. This was published as a short note by Goedkoop and Andresen (1955), along with a similar note on Al_2Th. This alloy came from the Philips Research Laboratories in Eindhoven, and was followed by related alloys, metal hydrides and new magnetic materials. In the early years Swedish crystallographers, also, took their problems to Kjeller (Aurivillius 1956).

Even before the JEEP reactor was put into operation an ingenious single-crystal instrument (figure 4.2) had been constructed and built by

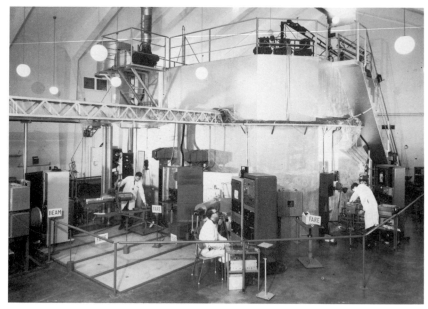

Figure 4.1 The JEEP reactor at Kjeller in 1959, with T Riste, O Steinsvoll and A F Andresen.

Figure 4.2 G Barstad's single-crystal diffractometer at Kjeller in 1955 (see 1957 *Rev. Sci. Instrum.* **28** 916).

G Barstad in the adjoining Norwegian Defense Research Establishment. This employed a macromodel of the crystal. By a parallel movement of the two, and subsequently of the counter arm, the Bragg condition could be brought to fulfilment by a geometric construction without performing any calculations. This instrument was used to collect the first neutron diffraction data on hexamethylenetetramine (Andresen 1957). In spite of all its ingenious features the instrument was difficult to modify for automatic operation. The instrument is now on display at the Norwegian Technical Museum in Oslo.

During his visit to Kjeller McReynolds (1954) constructed another simple instrument, which was modified later for moving the counter also in the vertical plane. This he used in cooperation with T Riste for studying the diffuse magnetic scattering away from the Bragg peaks. In order to obtain high enough intensity a white beam was employed. In the beginning interest was concentrated on magnetite, and this led to the first observation of magnetic critical fluctuations in a ferromagnet. Later haematite also was studied. Here it was possible to demonstrate for the first time the existence of a linear dispersion relation for spin waves in an antiferromagnet (Goedkoop and Riste 1960).

Kjeller, Petten and Delft. The American Atoms-for-Peace programme made enriched uranium available, and thus took away the *'raison d'être'* of JENER, which was formally dissolved in 1959. By that time the Dutch were building a materials-testing reactor at Petten. With this more powerful neutron source B O Loopstra and J Bergsma continued the work they had begun at Kjeller on elastic and inelastic scattering respectively. Powder diffraction was improved by using longer wavelengths and increased resolution (Loopstra 1966). At this time computers came to play an increasingly important role in spectrometer operation and data handling. With this in mind H M Rietveld joined the group at Petten. In cooperation with Loopstra, Rietveld (1969) developed the very powerful profile refinement method for treating powder neutron diffraction data. Independent of the JENER tradition a smaller reactor was built at Delft and used by J J van Loef and his students for neutron studies of molecular crystals, liquids and gases.

Through most of the 1960s a formal cooperation continued between Petten and Kjeller, where a more powerful heavy water reactor, JEEP II, equipped with a cold neutron source came into operation in 1967. This opened new possibilities for experimental research, and has since been exploited by T Riste and his group for the study of phase transitions, by K Otnes for the study of the dynamics of molecules, and by A F Andresen for further structure work. Parallel with this has gone methodological work, such as the implementation of flat and bent pyrolytic graphite crystals as monochromators.

Sweden

The first reactor in Sweden, R1 (figure 4.3), went critical on 13 July 1954. This was built on the premises of the Royal Institute of Technology close to the centre of Stockholm. The uranium came from CEA, France, and the heavy water from Norsk Hydro, Rjukan, Norway. Most of the employees were young, inexperienced, but enthusiastic people with a recent degree from the Technical University. The first year was spent in learning how to use the reactor, study its properties and starting some basic experiments. Production of counters was started, and methods developed for measuring neutron fluxes and standardising neutron sources.

The stimulus for starting neutron beam research was obtained through contacts with and visits to Brookhaven National Laboratory, in particular to the group of D J Hughes. After a visit in 1952–53 N-G Sjöstrand was stimulated to plan the building of a fast chopper for neutron cross section measurements, and later K-E Larsson spent the year 1955–56 at Brookhaven learning to use choppers in connection with time-of-flight equipment.

Although Brookhaven in those days was a real Mecca for neutron physicists, it was not always so easy to work there. Only one side of the reactor was declassified, and the time-of-flight equipment was situated in

Figure 4.3 The R1 reactor at Stockholm in 1960, showing the cold-neutron spectrometer with vertically moveable arm.

the classified area. It was only through the wholehearted backing of D J Hughes that Larsson was allowed to work there, but always with an armed policeman at his side. For those who know Karl-Erik personally it is no wonder that he developed a warm friendship with the whole Brookhaven police corps.

An important development in Brookhaven in those days was the use of a slow chopper for the study of phonon scattering from single crystals. On his return from Brookhaven Larsson, in cooperation with R Stedman, started putting together equipment for such studies including the installation of a cold neutron source in R1. In the course of 1957 a research group was formed consisting of S Holmryd, K Otnes, U Dahlborg and G Nilsson. Advantage was taken of the close cooperation established with Brookhaven to invite H Palevsky to Sweden where he spent nine months working with this group. Palevsky brought with him the idea of using the cold neutrons to try to verify the existence of Landau rotons in superfluid helium. This investigation, which turned out to be very successful, not only led to their verification, but also to the determination of their dispersion relation and its temperature dependence (Palevsky *et al* 1957).

Another research project taken up was the study of phonons and their lifetime in aluminium. Of particular interest was the observation that the scattering from liquid aluminium was similar to that of polycrystalline aluminium. This showed the existence of some sort of 'phonons' in the liquid state and led to the development of molecular hydrodynamics. At this time, 1959, a young theoretician A Sjölander joined the group, and his participation was of great value in the interpretation of the experimental results and helping to stimulate further research on liquids and gases.

After a visit to B N Brockhouse in Chalk River, R Stedman started building a triple-axis spectrometer. This instrument was, however, not put into use before the new, more powerful, reactor R2, which had been built at Studsvik south of Stockholm, became available in 1960. Here together with G Nilsson, Stedman built two more triple-axis spectrometers on which during the next few years a series of dispersion relations and frequency distributions in both metals and semiconductors were determined.

Neutron diffraction for structural studies was not started before 1963 at the R2 reactor. The initiative was taken by Professor G Hägg, University of Uppsala, who succeeded in obtaining the necessary funding for building two instruments, one powder and one single-crystal instrument. These were placed at the same channel using a tandem collimator and two lead monochromator crystals. To reduce the large flux of fast neutrons a substantial shielding was required. In addition to a collimator box filled with 6 mm diameter iron balls and flooded with water, the monochromator crystals were placed in a steel-walled water tank containing 18 tons of water.

The first powder diagrams were obtained in 1963, but the single-crystal instrument was not put into operation before 1964. This was a manually operated instrument, and the rate of data taking was typically 8 reflections per day. In 1965 an automatic diffractometer of the Hilger–Ferranti type programmable by punched paper tape was put into operation. It seldom functioned well, and was in 1969 replaced by a Hilger–Watts instrument which proved to be a much more reliable instrument. This was run through a PDP8 computer. The output was on paper tape, but after a few years this was changed to magnetic tape.

The personnel occupied with the running and improving of the instruments were in the beginning J Österlöf and O von Heidenstam, both with a background in X-ray diffraction. G Hägg remained head of the project until his retirement in 1969, when I Olovsson took over. In 1975 the instruments were rebuilt and the National Committee for Crystallography took over the responsibility. The first publication appeared in 1964 and was at Studsvik was carried out by Å Nilsson, R Liminga and I Olovsson on nised early on the importance of applying neutron diffraction to the study of heavy-metal compounds, and had in addition to powder diffraction research at Kjeller carried out a single-crystal study of Hg_2OCl_4 at Harwell. The first complete three-dimensional single-crystal structure determination at Studsvik was carried out by Å Nilsson, R Liminga and I Olovsson on $N_2H_5HC_2O_4$ (1968). A main emphasis has since been on hydrogen bonding. However, as the facilities were put at the disposal of all crystallographers in Sweden the problems have been taken from many different fields including ferroelectricity, phase transition and crystal chemistry.

Denmark
The Risø National Laboratory, situated 35 km west of Copenhagen, was founded in 1955 partly on the initiative of Niels Bohr. Its aim was to carry out research and technical development under the auspicies of the Danish Atomic Energy Commission. Its three reactors DR1, DR2 and DR3 came critical in rapid succession in the early part of the 1960s. Some neutron beam research was started at the DR2 reactor, but the main activity in this field had to wait until the higher flux of the DR3 reactor became available.

The initiative to build the first spectrometer was taken by the head of the Physics Department, O Koefoed-Hansen, in 1956, and two fresh graduates from the Technical University of Denmark, H Bjerrum Møller and J Schiellerup Petersen were given the assignment of building an instrument which could be used for measuring neutron scattering cross sections. A simple one-axis spectrometer was constructed and built by the mechanical workshop at Risø, and put up at a radial channel of DR2, a 5 MW light-water moderated reactor. It soon became clear that by this time most neutron cross sections had been well determined, and on return from a stay at Brookhaven National Laboratory, Bjerrum Møller in cooperation with W Koefoed started building Risø's first three-axis spectrometer TAS I. About the same time J Als-Nielsen and O W Dietrich took the initiative to convert the one-axis spectrometer into a two-axis spectrometer, TAS II. When the new reactor DR3 (a 10 MW, heavy-water moderated reactor of DIDO type) came critical in 1962 both instruments were put into operation at this reactor. This marks the onset of neutron scattering research at Risø and in Denmark.

After the first introductory investigations the work on TAS II was concentrated on the study of critical phenomena—active in this research were J Als-Nielsen and O W Dietrich supported by P A Lindgård who had joined the group as a theoretician. His close contact with W Marshall at Harwell proved very valuable in the interpretation of the experimental data. The first publications on phase transitions appeared in the mid 1960s and in the following years a series of important papers on critical phenomena was published.

TAS I, which was the only triple-axis spectrometer in operation at Risø up to 1967, was used for inelastic scattering investigations by H Bjerrum Møller, T Brun, J G Houmann and A R Mackintosh. Mackintosh started his activity in Denmark as a guest scientist in 1964, and took part in the first investigations on chromium. Later, as a professor at the University of Copenhagen, he acted as a consultant to the neutron physics group, and took an active part in the evaluation of data. It was on his initiative, and based on his international contacts, that an extensive research on the magnetic properties of rare-earth metals was started. This led to some of the first observations of magnetic excitations in a rare-earth metal, terbium (Bjerrum Møller et al 1967), and later to a series of publications both experimental and theoretical on rare-earth compounds.

As in the other Scandinavian countries the connections with Brookhaven National Laboratory proved to be extremely valuable. Not only did most of the scientists in these countries visit Brookhaven for shorter or longer periods, but experienced scientists from this laboratory worked on a guest appointment at the different laboratories in Scandinavia. For the development of neutron scattering at Risø the guest assignment of L Passell meant important new activities and increased international contacts. On his initiative the investigation of magnetic materials was started, a field which later developed into a main research activity of the neutron physics group.

In the period 1964–69 one of the new instruments at Risø was a time-of-flight diffractometer, used first for powders and later for single crystals. The instrument was part of a collaboration with B Buras and his co-workers at Swierk, Poland. Polish scientists visited Risø during these years, and at Risø, K Mikke in cooperation with B Lebech concentrated on developing the time-of-flight technique for single-crystal structure studies. It was, however, soon realised that the time-of-flight method was not an optimal one for studying magnetic structures of single crystals at a steady-state reactor.

Another instrument developed at the same time was the MARX-spectrometer (multi angle reflecting x-tal spectrometer) which J K Kjems constructed in cooperation with P A Reynolds and J W White from Oxford University. This is a triple-axis spectrometer incorporating a linear position-sensitive detector in the analysing part, and combines the resolution and focusing properties of the triple-axis spectrometer with the high data acquisition rate of a time-of-flight instrument.

In 1964 the Danish National Committee for Crystallography applied for, and got, 397 000 Danish kroner to put up a four-circle diffractometer at the DR3 reactor. A Hilger–Ferranti instrument with punched paper tape input was acquired. The instrument was put into operation in 1967 (figure 4.4) and, as for the similar instrument in Studsvik, a series of difficulties was encountered. These were, however, solved during the first year, and the instrument remained in operation until 1981 utilising during this time about 95% of the reactor running time. The project head was S E Rasmussen of the University of Aarhus and most of the scientists using the instrument in the beginning, like I Sötofte, M S Lehmann and F Krebs Larsen, were based at this university. The experimental work was taken care of on frequent travels to Risø, and by stationing an able technician, M H

Figure 4.4 F Krebs Larsen and M Lehmann with the 4-circle diffractometer at Risø in 1967.

Nielsen, at this laboratory to take care of the daily work. During the years huge amounts of paper tape were transported between Risø and Aarhus.

References
Andresen A F 1957 *Acta Crystallogr.* **10** 107
Aurivillius K 1956 *Acta Chem. Scand.* **10** 852
—— 1964 *Acta Chem. Scand.* **18** 1552
Bjerrum Møller H, Mogensen (Lindgard) P A, Gylden Houmann J C and Kowalska A 1965 *Symposium on Inelastic Scattering of Neutrons by Condensed Systems, BNL 940 (C-45) 139* See also Bjerrum Møller H, Gylden Houmann J C and Mackintosh A R 1967 *Phys. Rev. Lett.* **19** 312
Goedkoop J A and Andresen A F 1955 *Acta Crystallogr.* **8** 118
Goedkoop J A and Riste T 1960 *Nature* **185** 450
Loopstra B O 1966 *Nucl. Instrum. Methods* **44** 181
McReynolds A W and Riste T 1954 *Phys. Rev.* **95** 1161
Nilsson A, Liminga R and Olovsson I 1968 *Acta Chem. Scand.* **22** 719
Palevsky H, Otnes K, Larsson K-E, Pauli R and Stedman R 1957 *Phys. Rev.* **108** 1346

4.2 The First Experiments in France

B Jacrot† and D Cribier‡

†European Molecular Biology Laboratory, Grenoble, France
‡Centre d'Etudes Nucleaires de Saclay, France

In 1948 the first French reactor ZOE was completed at Fontenay aux Roses and delivered its first neutrons. Around this significant national achievement the Commissariat à l'Energie Atomique (CEA) assembled a group of young engineers to work on reactor physics. In contrast the situation of French academic science was then very bad after five years of almost total inactivity. Moreover the supervision of young scientists was often non-existent. One of the tasks of the CEA engineers was to measure cross sections of fissile materials. Another was the study of the thermalisation of neutrons. In order that the first task should go as quickly as possible two groups were working in parallel on the reactor EL2 which started to operate at Saclay in 1952. One group was doing spectrometry with Bragg reflection from crystals, while the other used time-of-flight methods. Curiosity led some of the young men involved in those studies to apply the

(Extrait des *Comptes rendus des séances de l'Académie des Sciences*.
t. 240, p. 745-747, séance du 14 février 1955.)

PHYSIQUE NUCLÉAIRE. — *Mesure de l'énergie de neutrons très lents après une diffusion inélastique par des polycristaux et des monocristaux*. Note de **M. Bernard Jacrot**, présentée par M. Frédéric Joliot.

On a effectué les premières mesures directes du spectre en énergie des neutrons de quelques millièmes d'électron-volt après diffusion inélastique par des substances mono- ou polycristallines. Ces mesures constituent un stade préparatoire à l'étude systématique par cette méthode du spectre de vibration d'un réseau cristallin.

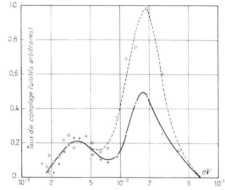

Fig. 1. — La courbe en trait plein est relative au cuivre polycristallin. La courbe en tirets est relative au monocristal de cuivre.

Figure 4.5 The observation of the first phonon.

techniques to what we call now solid state physics. It was quite natural for anyone familiar with the problems of neutron thermalisation to try to cope with the basic physics involved in that process. A paper by Peter Egelstaff led one of us to modify the chopper used for the cross section work so that it would be possible to measure the energy transfer during scattering. The first inelastic scattering by individual phonons was observed in 1954 (figure 4.5) and this was very soon followed by the analysis of the neutron scattering by iron in the neighbourhood of the Curie point. The activity in this last field of research was stimulated by personal contact with Van Hove, established on the occasion of a summer school at Les Houches, a place where we could get the training that we had not received from the university.

Almost immediately we had to build a new machine, for the first one was so noisy that it made mental concentration impossible and induced irreversible damage to our hearing. The electronics of this new machine provided 100 channels, a big improvement compared with the first one which had only 10. However the content of the channels still had to be read individually every hour and written down in the log book. The five years gap of the war had dramatic consequences on the development of French electronics whereas in England the war had rather been a stimulus for that technology.

The fact that our study of inelastic scattering started before any work on diffraction may be not unrelated to the existence of a rather strong group working with Jean Laval on phonons, whereas crystallography in France was certainly not at that time the best in the world. However in 1956 Pierre Meriel came to Saclay and, in association with Andre Herpin, modified the the diffraction instrument used for cross section measurements into a very crude but usable powder diffractometer. One had to be willing to spend days and nights around it, for the flux of the beam was unstable and no monitor was then available. Quite naturally the very strong French tradition in magnetism lead our colleagues to work on magnetic structures. The first work was on yttrium iron garnet (YIG), a material developed at Grenoble and whose atomic structure was already solved: this work was published in 1954. Many magnetic materials were subsequently investigated and one of them, Au_2Mn, gave peaks which were incompatible with the crystalline structure. At this time in 1959 Jacques Villain was looking for all theoretically possible magnetic structures in the case where the interactions with first and second neighbours are of opposite sign. One of the configurations which he predicted was a helical structure with a pitch incommensurable with the crystal lattice and this was found to account perfectly for the peaks observed experimentally with Au_2Mn. This was a nice example of cross fertilisation of experimentalists and theoreticians working together.

It was obvious that inelastic scattering with time-of-flight methods, which use long wavelength neutrons, would benefit enormously from any device which would enhance the flux of those neutrons. For anyone familiar with neutron thermalisation the concept of a cold source was an obvious one and indeed first, but unsuccessful, attempts to cool neutrons had already been made as early as 1951 with a neutron beam from the reactor ZOE at Fontenay. The successful demonstration of the cooling of neutrons by Peter Egelstaff prompted us to build a specially designed cold source using liquid hydrogen for the reactor EL3 which was then under construction (in fact as our source was too thick we had to use a mixture of hydrogen and deuterium). This installation went very smoothly without any problems; our success, which our followers have not always encountered, was due to two factors: first, there was very efficient participation by the inventive cryogenic laboratory of Professor Weil in Grenoble and, secondly, a very important factor was the intelligent approach to the safety problems by those responsible for the reactor. The cold source was ready as soon as the reactor went into operation and could then be used for a large programme on critical phenomena.

4.3 The History of Neutron Scattering in West Germany

H Dachs

Hahn-Meitner Institut für Kernforschung, Berlin

The start in Munich

In 1955 the treaties allowing West Germany to participate in peaceful nuclear research were signed. Politicians, administrators and physicists jumped at the opportunities which opened up. As early as 1957 the first German research reactor was put into operation in Garching near Munich. By this time neutron scattering was already a well established method. It was introduced to German physicists by Cliff Shull at the Munich meeting of the German Physical Society in 1956.

The development of neutron scattering in Munich was a rapid process, and fast processes have a tendency to be adiabatic. So in many respects research in Munich started from scratch, and it turned out that much pioneering work had to be done. At the reactor Tasso Springer, a PhD student, aroused the indignation of health physicists by a careless experimental set-up. For his measurements of the cross sections, he directed an unshielded neutron beam through the reactor hall. After a rebuke, he put a long copper tube in to prevent the beam from passing through. And, surprisingly enough, the intensity on his counter increased! He had caught the neutrons in the copper tube by total reflection; the neutron guide was discovered (Christ and Springer 1962). Tasso Springer was ready for his 'long march' through director chairs.

The reactor at Garching was linked to the Technical University Munich and was under the direction of Professor Maier-Leibnitz, a nuclear physicist of the famous Heidelberg school (figure 4.6). He had come to Munich only shortly before and had subsequently transferred to the field of neutrons. The new ideas he introduced while conducting a number of experiments were to characterise the second generation of neutron diffraction equipment. Figure 4.7 shows the 'atom egg' at Garching under construction in 1957.

In addition to total reflection in neutron guides another simple principle, that of back-scattering, was put forward to achieve very high resolution either in spectroscopy, for the measurement of very low energy transfer, or to determine accurate diffraction angles or line shapes (Alefeld 1966).

Gravity, which causes neutrons to fall, was used to measure scattering lengths. Only after neutrons have fallen some considerable distance are

Figure 4.6 H Maier-Leibnitz at Garching.

Figure 4.7 The 'atom egg' at Garching under construction in 1957.

they loaded with energy high enough to penetrate the surface of a piece of particular material, otherwise they are reflected totally. A collection of very precise values based on this principle is due to L Köster. The measurement of the scattering lengths was, as far as practical results are concerned, the most important Munich contribution, while otherwise the low power of the reactor (1 MW at the beginning) forced a concentration on methodological questions.

Furthermore, an attempt was made to build a neutron interferometer and to slow down the neutrons to a very low speed (A Steyerl).

The number of students starting work at the Garching reactor was considerable, as was the number of problems studied. And the influence of Munich was just as great and not confined only to Germany. Ideas, people and a unique language were exported. H Maier-Leibnitz had introduced the momentum space diagrams, a representation of scattering experiments which allowed an optimal experimental set-up to be inferred without resort to details of a spectrometer's design. The corresponding jargon spread with the Bavarians, and also the Bavarian greeting '*Gruss Gott*' which became a sort of radioactive tracer of the Munich diffusion. If heard in the reactor it was more than likely that the Garching Mafia had established themselves.

'The Grossforschungszentren' in Jülich and Karlsruhe

In the big nuclear research centres in Jülich and Karlsruhe development was not as rapid as in Munich. Because the completion of the reactors took a lot of time, the friendly assistance offered from Mol in Belgium and Würenlingen in Switzerland was called on initially. In Mol H Stiller started his work. At first he put water into the neutron beam—nasty people said he could not find anything else in the town of Jülich. Primary decisions are often far-reaching and in this case it was also to determine a major part of the later Jülich programme.

In 1968 the Jülich programme was widely extended. An Institute of materials research and neutron scattering was founded. Springer was appointed to Jülich and he brought with him several members of the Munich team, B Alefeld, T Heidemann, R Scherm and W Schmatz: the tradition of developing new methods was transferred to Jülich too. Being a 'Grossforschungsanlage' the Jülich equipment could be designed in a way that enabled the entire scientific community to make use of it, and ensured that it did not automatically crumble after the departure of the PhD students as it is customary with universities.

The ideas born and proved feasible at Munich developed into reliable and generally applicable methods. Jülich was of paramount importance as an intermediary which ensured the successful functioning of the ILL in Grenoble.

The flux of the reactors in Jülich was now satisfactory for a real research and measurement programme. A cold source ready for operation from 1969 opened up new vistas.

Professor Stiller's first measurements on water developed into systematic investigations of hydrogen bonds and protons in condensed matter in general.

Milestones were

the determination of the distribution of protons in the hydrogen bonds of KDP from incoherent scattering;

studies of different solid phases of methane performed by the experimentalist W Press in cooperation with the theoretician A Hüller;

the representation of the spatial distribution of protons by symmetry-adjusted functions, an important contribution to general structural crystallography.

But Jülich's most important success in this field was the evidence for tunnel-splitting by W Press, who demonstrated this with methane, and by B Alefeld, who used 4-methyl-pyridine. The improvement of high resolution spectrometers which are necessary for this kind of research had been achieved by Alefeld.

The second line of research at Jülich arose from disorder studies. On a suggestion from the AERE Harwell, Springer had already, when in Munich, used a beam of cold neutrons from a neutron guide to do experiments on small-angle scattering from dislocation lines in copper. In Jülich the type of small-angle apparatus with a long flight path was created and this is shown in figure 4.8. The famous D11 of the ILL in Grenoble, which is of the same type, was designed at Jülich, too. The considerations of how to improve small-angle scattering, as well as ideas on many other questions of disorder studies, are due to Professor Schmatz (Schmatz *et al* 1974). The development of the method of small-angle neutron scattering might be considered as one of the most important European contributions to the technique of neutron scattering. Small-angle scattering was widely used in Jülich:

Measurements on type II superconductors started in 1969 in Jülich after flux lines had been detected at Saclay.

The labelling of polymer chains by deuteration was discovered independently at almost the same time in Germany (C G Kirste at Mainz), England (G Wignall at ICI) and France. The first samples from Mainz and England were measured at Jülich in 1970.

Investigations on the characterisation of steels started in 1974. Quantitative results on the shape of segregations were derived from the small-angle scattering signal.

Figure 4.8 The small-angle scattering installation at the reactor in Jülich.

Contrary to the experience at Jülich, at Karlsruhe neutron scattering was started as an integral part of the reactor development programme. Here, cross sections that were important for the calculation of neutron diffusion in moderators were measured. Later, the cross section measurements on hydrogen turned out to be useful for the construction of the cold sources at Karlsruhe and Grenoble. The measurements on methane helped to produce a programme that was later successful at Jülich. Professor K H Beckurts was in charge of these studies. One member of his group was a young PhD student, W Gläser, who not only believed in his master but took an interest in what was going on around him. Professor Buckel's institute of low temperature physics was located in Karlsruhe. So Gläser began with studies of superconductors and hard materials. This led to early and important contributions on questions of electron–phonon coupling. The greatest success in this field was the confirmation of the 'Giant Kohn Anomaly' in the one-dimensional conductor $K_2Pt(CN)_4Br_{0.3}.3D_2O$ (Renker *et al* 1973).

Many materials which they wished to study were not available as single crystals, e.g. many A15 compounds. Therefore they confined themselves to the measurement of the densities of states using crystal powders. This method turned out to be useful for metals like Ca also.

All this time the research centre in Karlsruhe had been a generous host to visitors from universities. In later years it organised a visitors' service like a 'little-Grenoble'; that it worked so well is thanks to G Heger.

Crystallography institutes in the early years

In the field of crystallography the contacts with the American tradition were closer. After C G Shull's talk in Munich in 1956, Professor Maier-Leibnitz proposed to the crystallographer Professor Menzer that he should perform measurements at the Garching reactor. The consequence was that Professor Menzer sent me to MIT and Brookhaven in 1958. Back at Munich I started studies on hydroxides and magnetic structures and tried to improve the method of diffractometry with curved monochromators. Later, under the supervision of Professor Jagodzinski the programme comprised certain problems of disorder studies and incommensurate phases.

Munich spawned two more groups. In 1967 I went to Tübingen. The Tübingen Institute of Crystallography used the reactor in Karlsruhe. Phase transitions were studied and, with the help of the Tübingen chemists, one- and two-dimensional magnetic systems. The consequent continuation of studies of the one-dimensional ferromagnet $CsNiF_3$ finally led to the detection of solitons (Kjems and Steiner 1978). But these studies on solitons were done after I had already moved to the new reactor at Berlin and taken half of the Tübingen group with me. The Tübingen institute was then taken over by Professor Prandl, another former member of the Munich institute, and he extended the neutron programme even further, for example by studies of amorphous materials.

In 1957 Professor Wölfel began with the construction of a neutron spectrometer in Darmstadt which was transferred to Karlsruhe because the small reactor in Frankfurt had turned out to be too weak. The studies in Darmstadt were concerned with the determination of electron densities with X-rays. The neutron investigation resulted in an independent determination of the temperature factors. Later on magnetic studies were added. Measurement of temperature factors was continued by Müllner from the Frankfurt reactor centre.

In the early 1960s the Mineralogical Institute in Bonn began constructing a neutron diffractometer which was set up at Jülich. From 1969 onward the programme was developed by Professor Will. He had studied at Brookhaven with G Shirane and R Nathans and came to Bonn via Darmstadt. His studies were concerned with magnetic structures of rare-earth compounds, especially rare-earth borides.

The desire to clarify the function of water in minerals led the Mineralogical Institute in Frankfurt to use neutrons. The study of zeolites (Bartl) was an important part of it. Felspars were integratated into the programme by Professor Korekawa and H Fuess added X–N synthesis and spin-density determinations. The Frankfurt group also worked at Karlsruhe.

The period since the ILL went into operation

From the second half of the 1960s many German groups took part in the development of the ILL. After the Grenoble high-flux reactor had been put into operation, the number, kind and quality of investigations with neutrons increased vastly. The extension of the university support run under the name of 'nukleare Festkörperforschung' enabled many new groups—at the beginning of the 1980s there were about thirty of them—to work with neutrons.

While studies of inelastic neutron scattering had previously been restricted to groups working at the Grossforschungszentren, university groups were now able to take part in this field of research. As new groups Aachen, Bayreuth, Mainz and Würzburg are to be mentioned; Frankfurt, Munich and Tübingen are still most important among the old ones.

German chemists discovered neutron scattering, too. Extensive programmes are now being run at the institutes of inorganic chemistry in Aachen, Hamburg and Munich, as well as at the Institute of Physical Chemistry in Aachen and at the Max Planck Institute in Stuttgart.

The work of H Stuhrmann was responsible for a new direction in biological research. He explored the possibilities of contrast variation in the embedding of biological macromolecules in different light–heavy water mixtures. W Hoppe contributed the idea of determining the morphology of assemblies of proteins by triangulation. In other words, the different proteins were deuterated one after another, and their distance was determined by small-angle scattering. This method is now being used by K Nierhaus at the MPI of molecular genetics in Berlin to study the shape of the large subunit of ribosome, a project that will drag on for years.

In the last ten years several other research reactors have gained in importance. The new reactor in Berlin has already been mentioned. The upgrading of this reactor is now being pursued by A Axmann. The programme in Brunswick—well known as the place from which the skilled experimentalist O Schärpf came—has been able to extend considerably (e.g. investigations in polarised nuclei) after the appointment of Professor Scherm.

The reactor station in Geesthacht near Hamburg has recruited M Wagner, a metal physicist of the Göttingen school. He is now widening the use of his reactor and making it a focus for metal research.

On the other hand the reactor in Karlsruhe was closed down in 1981. Professor Gläser transferred to Munich and Professor Schmatz, who is in charge of the Karlsruhe group, tried to secure continuation of the Karlsruhe research programme by moving instruments to French reactors.

Acknowledgment

Many thanks to my German colleagues who supported me with historical material and to K Diederichsen who translated the German manuscript.

References

Alefeld B 1966 *Bayerische Akademie der Wissenschaften, Mathematisch-Naturwissenschaftliche Klasse, Sonderdruck 11*
Christ J and Springer T 1962 *Nukleonik* **4** 23
Kjems J K and Steiner M 1978 *Phys. Rev. Lett.* **41** 1137
Renker B, Rietschel H, Pintschovius L, Glaser W, Brüesch P, Kuse D and Rice M J 1973 *Phys. Rev. Lett.* **30** 1144
Schmatz W, Springer T, Schelten J and Ibel I 1974 *J. Appl. Crystallogr.* **7** 96

4.4 'Start-up' in Italy

G Caglioti

Istituto di Ingegneria Nucleare, CESNEF, Milan, Italy

Since the time of James Chadwick, neutrons have been popular among Italian physicists. As Emilio Segre recollects in the *Enrico Fermi— Note e memorie* (collected papers), immediately after the discovery of the neutron, Fermi realised that the research on artificial radioactivity, undertaken by I Curie and F Joliot using charged particles as projectiles, could be expanded tremendously by using neutrons instead. In his Nobel speech on 10 December 1938, Fermi states: by neutron bombardment 'a systematic investigation of the behaviour of the elements throughout the periodic table was carried out by myself with the help of several collaborators, namely Amaldi, D'Agostino, Pontecorvo, Rasetti and Segré.'

In the middle 1940s, as soon as 'intense' neutron fluxes became available at the Argonne pile, Fermi's interests inevitably focused on aspects of the physics of neutrons, and physics by neutrons, which had been assigned low priority in the drive towards wartime objectives.

When in 1955, a year after the death of Fermi, the Italian Government decided to install a CP-5 type research reactor at Ispra, Edoardo Amaldi, the then vice-president of what was then the Comitato Nazionale per le Ricerche Nucleari (CNRN), could not remain indifferent to the beautiful physics which was being developed using reactor neutrons at the Argonne, Oak Ridge, Brookhaven, Harwell and Chalk River laboratories. In 1957, while the Ispra reactor was being constructed, Amaldi decided to form a group with the objective of building a crystal spectrometer to explore the structure and properties of condensed matter.

I cannot refrain here from disclosing a personal recollection. When Amaldi, my Maestro, invited me to contribute to the research programme which he was heartily supporting, I happened to be engaged, both as a CNRN researcher and as a resident research associate with the physics division of the Argonne National Laboratory, in the experimental determination of the half-life of the neutron decay, using a diffusion cloud chamber. At that time, while I was there, I was fascinated by the American way of work and life, to the point at which I had reached the decision to emigrate to the United States: but Amaldi's determination was inflexible!

I sailed back to Italy. Once in Rome, I shared an office in the Institute of Physics of the University, with my friends and colleagues Antonio Paoletti (who has kindly helped me to collect these memoirs) and Francesco Paolo Ricci. Around the end of 1957 the group was strengthened by Marcello Zocchi and Antonio Santoro, two chemical crystallographers, who added their scientific interests to our research programme. Soon we realised that a mechanical engineer was needed to help us to write the specifications for the spectrometer, to design it, and to supervise its construction by

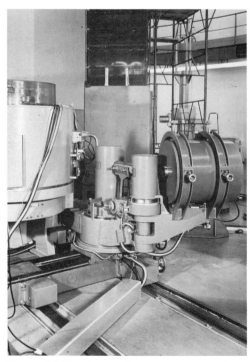

Figure 4.9 The three-axis spectrometer at Ispra.

Nuova San Giorgio in Genova and its installation and alignment at the Ispra reactor beam. Our need was recognised as sound, and Francesco Marsili was promptly employed by the CNRN. His stimuli and his work during the early years of neutron diffraction in Italy proved invaluable. We benefited also from the help of professional electronic physicists, like Umberto Pellegrini and Elio De Agostino.

After September 1957, the group at Rome devoted several months to a systematic scanning of the bibliography, and to a critical analysis of the effect of the collimator widths on the resolution and luminosity of crystal spectrometers. This analysis was originally undertaken both to optimise the performance of the spectrometer, while dreaming of it, and to answer insistent requests by Marsili. Our results were confirmed later by C G Shull and B O Loopstra, and in the early 1960s they were extended to the case of the three-axis spectrometer. I am told that our work of those years is still being utilised as a basis for the deconvolution of the intrinsic line widths from the measured intensity of the scattered neutrons in experiments of conventional diffraction, elastic diffraction† and inelastic scattering.

While the design and construction of the spectrometer and the installation of the Ispra neutron source proceeded, the physicists of the group were offered by the CNRN the privilege of working for periods of the order of one year in the most important North American laboratories.

Paoletti went to Brookhaven, Caglioti to Chalk River and Ricci to the MIT.

At Brookhaven, Paoletti worked with Robert Nathans in the field of magnetism. At the polarised neutron spectrometer Paoletti collaborated initially on the determination of the distribution of the magnetic moments of iron on the different sites of the ordered alloy Fe Al. Subsequently, again with Nathans, he determined the magnetic form-factor of the 3d

---

† Elastic diffraction, performed for example by a three-axis spectrometer whose analysing crystal is set at the impinging neutron energy, is conceptually different from conventional (two-axis) diffraction.

In conventional diffraction, contributions are collected from all diffracted neutrons, irrespective of their energies, so that every single neutron takes a sort of instantaneous picture of the system. In elastic diffraction, on the other hand, contributions are taken only from outgoing diffracted neutrons whose energies coincide with that of the incident neutron within the experimental energy resolution, ΔE, of the (three-axis) spectrometer: every single neutron thus takes a sort of time exposure of the system over a time t of the order of $8h \ln 2/\Delta E$, where h is Planck's constant.

Although conventional and elastic diffraction are basically different (especially when looking at liquids or hydrogeneous solid substances for example), a sloppy semantic superficiality still induces professional researchers to refer to *conventional* diffraction of neutrons or X-rays in terms of *elastic* diffraction. The present occasion is perhaps the last important one for me to fight such a misconception!

electrons of cubic cobalt, and discovered an important deviation of the distribution of the magnetic electrons from spherical symmetry. This work marked the beginning of an intense experimental and theoretival activity on the detailed distribution of the magnetisation density in solids.

At Chalk River, Caglioti worked with Bert Brockhouse in the field of the dynamics of atoms in crystals. The experimental work on neutron inelastic scattering in 1959 had just moved from NRX to the more intense NRU reactor, and the 'constant Q' method had just been invented. Brockhouse was a real Cicerone to all the subtleties of reciprocal space–time, for all the researchers converging to his group from Canada and from all over the world. With Brockhouse, Arase, Rao, Sakamoto, Sinclair and Woods we measured the phonon dispersion relations of lead at several temperatures. We thus derived the interplanar force constants and, furthermore, we had access to the intricacies of the electron–phonon interactions responsible for the Kohn anomalies related to the caliper dimensions of the Fermi surface in lead.

At Cambridge, Ricci worked with Shull on theoretical problems and experimental methodologies: unfortunately the MIT reactor was not ready yet; but nevertheless all of us in the group benefited from the experience Ricci gained from Shull at the time.

Meanwhile, back home, the situation was changing. Ispra became a joint research centre of the CEE, and the then Comitato Nazionale per l'Energia Nucleare (formerly CNRN, now ENEA), had decided to build its centre for nuclear research near Rome at Casaccia, and to install in it a Triga Mark II research reactor.

A bifurcation occurred: as originally planned, Caglioti, Ricci, Santoro and Zocchi moved to the CNEN Laboratory of the Euratom Centre of Ispra, while Paoletti and Marsili installed a polarised neutron spectrometer at the Triga source of Casaccia.

In 1960 Italy was a leading country on the international map of neutron spectrometry: the first European three-axis neutron spectrometer for inelastic scattering and the first European polarised neutron spectrometer were operating in Ispra and in Casaccia respectively. Ispra and Casacccia attracted a number of distinguished visitors from abroad. The friends we made at that time have helped our collaborators, directly or indirectly, to preserve the neutron tradition in Italy: like vestals, Franco Rustichelli and Filippo Menzinger are still blowing the fire of neutron research.

In the 1960s, our research programmes were naturally being developed in accordance with the specific competence previously acquired. In Ispra phonon dispersion relations for acoustic and optic modes in zinc at several temperatures were determined, and several studies on crystallographic and liquid systems (bromine, zinc and gallium) were performed using both conventional and elastic diffraction. In Casaccia an ample programme was developed on the distribution of magnetic moments in ordered ferro-magnetic alloys (Co–Pt, Fe–Pd, Fe–Si, Mn–Ni). Furthermore other neut-ron scattering groups were being formed, around other research reactors, namely at the Centro Studi Nucleari Enrico Fermi of the Politecnico di Milano, at the SORIN in Saluggia and later at the CAMEN in Pietrogrado, near Pisa.

Suddenly, during midsummer 1963, a major shock—*il caso Ippolito†*—shook the CNEN. That remains an enigma, for many of us marking the beginning of the end of an unrepeatable and stimulating scientific and human experience.

† *Editor's note.* Professor Felice Ippolito had, from the beginning, been the general secretary of what was in 1963 the Comitato Nazionale per l'Energia Nucleare. He became a victim of political intrigue and, after incrimination, spent a few years in gaol. Afterwards, Professor Ippolito was elected senator in the Italian Parliament and reinstated in his position of University Professor in Naples. Currently he is a member of the European Parliament.

4.5 A Dialogue from the Soviet Union

R P Ozerov† and A Yu Rumyantsev‡

†*Mendeleev Institute of Chemical Technology, Moscow, USSR*
‡*Kurchatov Institute of Atomic Energy, Moscow, USSR*

When the editor invited me (RPO) to participate in preparing this collec-tion of papers, I thought it would be worthwhile to outline the history of our research in the USSR in the form of a discussion between two persons. One of them is a specialist who has been at the foundations of the work and has worked with neutrons for many years. The other person is a relatively young man who, nevertheless, is deeply involved in these studies. I have chosen A Yu Rumyantsev from the Kurchatov Institute of Atomic Energy (IAE) to be the second person. Below follows the text of our conversation, which meets well-enough the objectives set by the editor.

AR It is well known that neutron diffraction in the USSR (where we also call it neutronography) began from your reviews, where you gave the main principles of the use of neutron diffraction in crystal-structure analysis. (These reviews were published in the journal 'Uspekhi Fizicheskikh Nauk', UFN, in the years 1949 to 1952). How did you come to enter this field of research?

RO It was long ago, about a couple of years after the Second World War, when I was a student of the Institute of Physical Engineering (which had not yet been so named). At the same time I was working as a technician in the X-ray laboratory headed by G S Zhdanov. One day German Stepanovich called my attention to a new physical phenomenon which could be used for studying crystal structures, namely the diffraction of neutrons. He recommended me to give a talk on this subject at the student research circle. I started working with great enthusiasm, gave my lecture and then prepared the review which was published later in the UFN on the initiative of E V Shpolsky who was the Chief Editor of this Journal at that time (see Ozerov 1949, 1951, 1952).

AR And when did experimental work begin?

RO Unfortunately, due to certain circumstances, I had to stop working on neutron diffraction for several years and to deal with vanadium-bronze synthesis and X-ray analysis. For this reason, the initial neutronography experiments were carried out by my future colleagues. These investigations were performed at the reactor of the First Atomic Power Station (in Obninsk) and at the reactor of the Institute of Theoretical and Experimental Physics (ITEP) of the Academy of Sciences of the USSR (in Moscow). These reactors were designed mainly for studies in the field of nuclear energy and nuclear physics. In Obninsk the research group, headed by N V Ageev and V N Bykov, carried out studies of a number of interstitial phases (Bykov *et al* 1957) and then started systematic investigation of the magnetic structure of chromium (Ageev *et al* 1958, Bykov *et al* 1959). It should be pointed out that this research team discovered satellites in the neutron scattering by chromium single crystals, and attributed them to magnetic and atomic incommensuration, independently and a little bit earlier than Corliss and Hastings. Studies of ordering and defects in the macrostructure of Fe–Ni alloys were carried out at the ITEP with Abov's spectrometer by Lyashchenko in 1957 and 1958. At the same Institute I I Yamzin and Yu Z Nozik studied the atomic and magnetic structure of manganese ferrites.

The IRT reactor at the Kurchatov Institute of Atomic Energy was one of the first reactors mainly designed for studies with neutron beams. I started my experimental work with colleagues at this reactor in 1958 as a guest scientist. We had to begin our work practically from scratch, since at that time there were no commercially available spectrometers, no end-window counters and no monochromators. Like E O Wollan, we used parts from an X-ray diffractometer.

However, in order not to make the detector arm too heavy, we developed a global shield which completely covered the whole of the spectrometer. We used the same principle later on, when we moved to our own WWRC water-moderated reactor in Obninsk, at the affiliated institute of the Karpov Institute of Physical Chemistry. At that time we installed an arc on a diffractometer arm, so that it could be used for single-crystal diffractometry by an orthogonal beam method. With this instrument we carried out all our structural work, for example studies of hydrates and the atomic and magnetic structures of double oxides.

At the end of the 1950s host IAE specialists advanced neutron investigations with the beams neighbouring ours. M G Zemlyanov and N A Chernoplekov developed instruments for studying the inelastic neutron scattering. Their pioneer works on studies of the vanadium phonon spectrum and of some disordered systems are well known. The first neutron crystal-spectrometer for measuring neutron cross sections was built at the IAE by Yu Ya Konakhovich and I S Panasyuk (Konakhovich and Panasyuk 1959). With the use of this installation the neutron diffraction investigations were intensively developed during the late 1950s and early 1960s. When V A Somenkov came to the IAE in 1960, the studies of structure and phase transitions in interstitial systems began. In the middle of the 1960s V A Somenkov and S M Shilstein carried out the first experiments on studying neutron-optical effects by means of neutron scattering by ideal crystals.

At the same time, on the initiative of I V Kurchatov and A P Alexandrov research centres based on nuclear reactors were established in many cities of the country. Notably, at this time in the A F Ioffe Institute of Technical Physics a new method was developed for producing polarised neutrons by reflection from polarised mirrors. Experiments on magnetic phenomena in ferromagnets were performed there, with this method, by G M Drabkin's team in Gatchina. These experiments stimulated theoretical research in this field by S V Maleev and others. Now all this activity is concentrated at the Institute of Nuclear Physics, which emerged from the Ioffe Institute. Traditional neutron-diffraction investigations of magnetic materials were started there by V P Plakhtiy .

Such centres have been established in many capital cities of Soviet republics. Subsequently, research reactors have also been built in many socialist countries with the assistance of the Soviet Union.

Returning to the I V Kurchatov Institute of Atomic Energy, one should recall that at this Institute diffraction experiments by the time-of-flight method were started in 1970, using the linear accelerator. The use of accelerators in neutron scattering experiments was initiated much earlier by M I Pevzner and V I Mostovoi. It is worth noting that the idea of applying the time-of-flight method to neutron diffraction had 'hovered in the air' for some time. In particular, this idea had been advanced by V V Safronov who was then working at the IAE but, unfortunately, experiments had not then begun.

AR Now on the basis of the 'Fakel' linear accelerator installations are built which allow the study of both neutron diffraction and inelastic scattering by the time-of-flight method.

RO That is right. The Soviet Union has made a great contribution to the development and use of the time-of-flight method in general. These achievements have been promoted by using the pulsed reactors IBR-1 and IBR-2 at the Joint Institute of Nuclear Research (JINR) in Dubna. The IBR-1 reactor was put into operation in 1960. Up to that time the research team, headed by I M Franck and F L Shapiro, was formed in the Neutron Physics Laboratory. Scientists from socialist countries also took part in the work with the IBR†. A detailed review of the major features and basic results obtained during the first ten years of operation of the IBR was given by I M Franck (1972). To tell the truth, I cannot describe it in a more interesting or elegant way than he did.

Nowadays a new and more powerful IBR-2 reactor is in operation. A great number of neutron scattering experiments have been carried out on this installation under the general supervision of I M Franck and Yu M Ostanevich. This work includes experiments in the field of biology (small-angle scattering), texture, molecular excitations, quantum liquids, etc.

One of the advantages of the time-of-flight method is the possibility of combining the diffraction unit with pulsed effects on a specimen. Thus, for example, pulsed magnetic fields have permitted field strengths up to about 120 kOe to be achieved and used in studies of the magnetic structure of haematite: such fields still remain the largest attained (Levitin et al 1969, Antsupov et al 1971).

The advantages and shortcomings of various pulsed neutron sources could be discussed at length but it's quite obvious that a pulsed reactor of the IBR type can be used with great success in a wide range of experiments.

The Laboratory of Neutron Physics at the JINR greatly stimulated the development of neutron studies in socialist countries. At regular workshops I had the pleasure of getting acquainted with leading specialists in neutron scattering—J Yanik (Poland), D Bally (Romania), K Hennig (DDR), J Natkants (Poland), D Zippel (DDR), A Byoreck (Poland), N Kroo (Hungary), and many others.

† A photograph taken at Dubna in 1965 appears as figure 4.10.

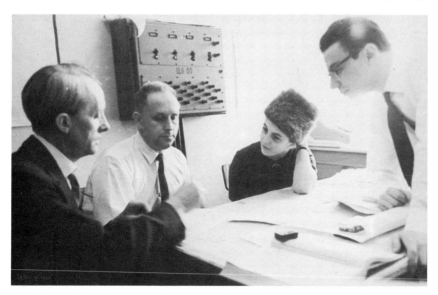

Figure 4.10 Discussing time-of-flight measurements of $BiFeO_3$ on IBR-1 at Dubna, 1965 (left-to-right, G E Bacon, R P Ozerov, I Sosnowska and J Sosnowski).

AR   Your name has been frequently associated with Yu A Izyumov's name. What is the field of your collaboration?

RO   First of all, we are simply good friends. Both of us were born in Sverdlovsk at a distance of one kilometre from one another, with a difference of 7 years in age, but fate brought us together considerably later. We got to know each other in 1962 on the Enissey river, on board the motor vessel 'A Matrosov', where a conference on solid state phyiscs took place. At that time Yu A Izyumov had only begun studies of neutron scattering in magnetic materials. Nevertheless we agreed at once to write a book on magnetic neutron diffraction, and it was indeed done. This book was published about 20 years ago in the USSR.

Later Izyumov turned himself to other problems of solid state theory, but recently I succeeded in attracting him to literary activity again. With his active participation we have issued a three-volume monograph *Neutrons and Solids*, which has reflected the modern situation in structural and magnetic neutronography and in neutron spectroscopy. Unfortunately, I have not done any scientific work in collaboration with him. However, recently at the Institute of Metal Physics of the Academy of Sciences of the USSR in Sverdlovsk an active collective, headed by S K Sidorov, has been formed. The collaboration between this team and Izyumov and his colleagues is, undoubtedly, for the benefit of both experimenters and theoreticians.

I believe the Kurchatov IAE may serve as a brilliant example of fruitful cooperation between experimenters and theoreticians, may it not?

AR Yes that is true. Solid-state studies with neutron scattering have been carried out at the Kurchatov IAE with close cooperation between experimenters and theorists from the very beginning. A great contribution to the development of studies of lattice dynamics had been made at an early stage by I I Gurevich, Yu M Kagan and L V Tarasov, and later on by E G Brovman. The joint work of theorists and experimenters is continuing now. It has contributed much to the discovery and systematic study of quasi-local and local levels in the phonon spectra of solid solutions, to considerable progress in the development of the electron theory of metals, and to persistent study of the effect of electrons on the fine structure of phonon spectra, etc.

We, however, have got too deep into theory; it is time to return to the main subject—to the history.

RO Just a moment, please! Before completing this subject, I would like to ask you about the perspectives of neutron studies in the Soviet Union.

AR At present, the neutron scattering work at the IAE is carried out with a modified IRT reactor, which is labelled IR-8. This installation is equipped with mono- and polycrystal diffractometers, two three-axis spectrometers and a time-of-flight spectrometer with a cold source (the 'Fakel' installations should be added to this list).

We associate our hopes with a new 'Peak' reactor installed at the Gatchina Institute of Nuclear Physics, of the Academy of Sciences of the USSR, and with the 'Meson Factory' near Moscow.

RO I would like to note that a very qualified team of scientists exists at the IAE (You, dear A Yu, are a member of this team). This team states and successfully solves many interesting and important problems.

Well, now we return to the history.

AR Yes. The conditions of work in the 1950s differ drastically from those existing nowadays. What are your impressions of that time? How did you meet your foreign colleagues for the first time?

RO Our work was really very interesting then. The fact is that only 25 to 30 people in the world were then working in the field of neutron diffraction. They often published their results in laboratory reports and it was well known what problems a given scientist dealt with, what were his results and, even, what were his future plans. The directions of work occasionally coincided, and it was interesting at that time (and even now) to compare various approaches and results. We, also, have sometimes been involved in such a competition.

All of us at some time have learned from G E Bacon's book

Figure 4.11 A party of crystallographers from the 1966 Moscow conference of the IUCr photographed at Obninsk. Back two rows, from left: S V Kiselev, G I Scherba,?——?, A S Boravik-Romanov, S P Sorovjev, E F Bertaut, L M Corliss, R P Ozerov, G E Bacon, S C Abrahams, K Burtsev, A Delapalme, R Nathans, S W Peterson, W Cochran. Front two rows: D Bally, P Meriel, P J Brown, K Knox, J B Forsyth, G H Lander, D V Dudarev, N Sokolov, M Betzl, I Olovsson, H A Levy.

Neutron Diffraction. So, it was especially interesting to get to know the 'living' Bacon. I well remember our first meeting at a railway station in Leningrad where he had come to the Federov Anniversary Session in 1959. After this we have met many times, both just ourselves and also with our families. To my pleasure, this outstanding scientist was found to be a gentle and sociable person. The same is true for all members of his family. During the next 3 to 5 years I got to know almost all people who were concerned with neutron scattering at that time. I still have friendly relations with many of them. This has been contributed to in a considerable degree by the Congress of the International Union of Crystallography in Moscow in 1966. Many foreign scientists came to our capital city, and we did our best to show them as many things as possible—the country, our institutions†, etc. It is a great pleasure and benefit for me to visit foreign institutes as well.

Now, a great number of new scientists have entered neutron scattering. The Neutron Scattering Conference in 1982 (the Yamada

† Photographs taken at Obninsk on this occasion appear as figures 4.11 and 4.12.

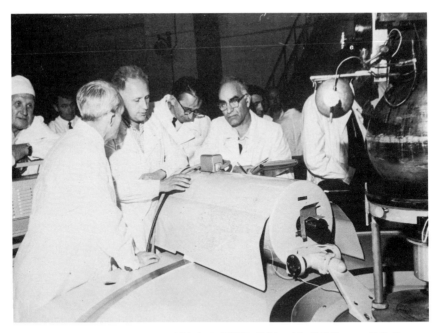

Figure 4.12 A diffractometer at Obininsk USSR 1966, with G E Bacon, R P Ozerov and H A Levy.

conference, Hakone, Japan) has shown that it is scarcely possible to know all your colleagues personally. This is also true as far as the scientists of a single country are concerned.

In conclusion, I would like to ask you one last question. Neutron diffraction is not the simplest way of earning your money and your reputation in science. Are you not sorry to be occupied with this difficult task?

AR I should mention two things in this connection. First, I am glad to deal with Science. Our teacher Academician L A Arzimovich joked that to work in science is the best way to satisfy one's own curiosity at the expense of the State. That is what I (and many of my colleagues) are doing. Secondly, hard work is indeed necessary to obtain experimental results using neutron scattering methods in solid state physics and, especially using inelastic scattering which is my interest now. It involves both physical and intellectual loads. Nevertheless, I have been busy with this work for 16 years already, and I think that if I now had to make a choice once again, I would behave in the same way as 16 years ago. I believe that I was lucky to work in this difficult and fascinating field of physics.

References

Ageev N V, Bykov V N and Trapeznekov V A 1958 *Izv. Acad. Nauk SSSR*
556

Antsupov P S, Voskanjan R A, Levitin R Z, Nyziol S, Nitz V V, Ozerov R P, Pak
Gvan O and Shafran S 1971 *Sov. Phys. Solid State* **13** 44

Bykov V N, Golovkin V S, Ageev N V, Levdik V A and Vinogradov S I 1959 *Dokl.
Akad. Nauk, SSSR* **128** 1153.

Bykov V N, Vinogradov S I, Levdik V A and Golovkin V S 1957 *Kristallographiya*
2 634

Franck I M 1972 *Problemy fyziky elementarnych tchastits i atomnogo yadra* **2** 806

Konachovich Yu Ye and Ponasjuk I S 1959 *Pribory i technika eksperimenta* **3** 26

Levitin R Z, Nitz V V, Nyziol S and Ozerov R P 1969 *Solid State Commun.* **7** 1665

Ozerov R P 1949 *Usp. Fiz. Nauk.* **38** 413

——1951 *Usp. Fiz. Nauk.* **45** 481

——1952 *Usp. Fiz. Nauk.* **47** 445

4.6 Early Years of Neutron Diffraction Work in India

P K Iyengar

Bhabha Atomic Research Centre, India

The first Indian nuclear research reactor APSARA became critical in 1956. Not surprisingly, the earliest neutron experiments were on measurement of spectra from the reactor using a slow chopper by Ramanna *et al* (1958) and with the help of a single-crystal spectrometer by Duggal *et al* (1961). Duggal built the first neutron crystal spectrometer in 1957. This was a simple device in which a well shielded BF_3 counter was manually rotated on wheels about the axis of a single-crystal monochromator. The crystal rotated independently and was adjusted optically. Surprisingly the measurements showed the effects of multiple Bragg scattering in the monochromator—known as the Renninger effect. This was published almost concurrently with similar work done elsewhere. The work on total cross sections as a function of energy using this single-crystal spectrometer, with gold foil and calibrated Pyrex filters and, later, a velocity selector to correct for higher-order effects, continued till about the middle of 1961.

I spent fourteen months in Chalk River, Canada, from the winter of 1956 to work with Brockhouse. On returning to India in May 1958 my task was to organise neutron work with the reactor APSARA and plan for the new heavy-water reactor being built at that time. My work started with the construction of a diffractometer based on a design similar to that at Chalk River. Fortunately, it was possible to get a gun-mount as a disposal item for a very small price. The diffractometer to be built on this as the base was taken up for fabrication in the divisional workshop. In August 1958, Satya Murthy and Dasannacharya joined me in this venture after graduating from the Training School of our establishment. Along with the main spectrometer the counting set-up and the automation were also fabricated in parallel within the small group. BF_3 counters had already been developed at Trombay for various other uses and came in handy. This was the first automatic spectrometer developed at Trombay.

By the year 1959 the spectrometer was taking shape to be installed at a beam tube. A lead single crystal grown by me at Chalk River was mounted on the central table of an ordinary classroom optical spectroscope to serve as the monochromator. This was accommodated in a recessed hole in the reactor shielding and a wall of concrete with an exit collimator provided the shielding (figure 4.13). The use of vacuum-tube pre-amplifiers and amplifiers picking up sparks from the electromagnetic relays and uni-selector switches of the automation circuits was a problem. As usual the physicists had to solve this in unconventional ways. How different it is now—with computers inside reactor halls! It was exciting to see the first results of a diffraction pattern from powder samples of aluminium—also to see the effect of cold rolling when a sheet was substituted. The reactor power had by now gone up to about 300 kW and it was not very difficult to see the Bragg peaks from thick powder specimens of various alloy samples.

The first samples chosen were alloys of iron with aluminium and tin. Japanese workers had suggested from magnetic measurements that $FeSn_2$ might be an antiferromagnet with a transition temperature above room temperature. This was a welcome substance for study since we were not yet equipped with a cryostat to go to low temperatures. An initial run at room temperature showed all the nuclear peaks but no extra reflection which we were hoping to see. However, the points near about $2\theta = 11°$ always showed a noisy behaviour. Adding several runs with a tighter collimation showed up the (100) reflection and a magnetic structure was derived from this. The work was reported at the International Conference on Magnetism in 1961 at Kyoto. This gave us confidence that, at last, neutron diffraction could also be done in a developing country.

As the construction of this spectrometer was progressing, another diffractometer ordered from John Curran & Co. in England arrived in 1959. With my interest and background in phonon work I decided to convert this diffractometer to a triple-axis spectrometer (Iyengar *et al* 1961). A simple goniometer was made in our workshop to change the crystal orientation of

Figure 4.13 The first neutron diffractometer at the APSARA reactor, 1959.

the Fe crystal which was pencil-shaped, being some three eighths of an inch in diameter and about two inches long. A tank of water with a tube for bringing out the monochromatic beam was used as the monochromator shield. After the usual calibration runs, energy analysis of incoherently scattered neutrons from a vanadium plate was made and the intensities gave hopes of being able to observe phonons. Then followed some weeks of hectic activity during June–July 1960, for it was certainly not easy to observe them. (Monsoon is not the best time of the year for electronic equipment.) Eventually we did manage to observe a few phonons and this was reported at the first IAEA meeting (Iyengar *et al* 1961). On the diffractometer, in the meantime, some diffraction experiments were carried out on water and heavy water, together with measurements of average energy-transfer using gold foils.

It was clear by the late 1950s that neutrons would play an important role in research in solid state physics and it was, therefore, decided to launch a broad neutron scattering programme with the new reactor at Trombay in preference to nuclear physics. The strength of this group was being built up continuously. Venkataraman and Usha, who had gone to Chalk River to work with Peter Egelstaff, returned in April 1959. They started designing a rotating crystal spectrometer and building a hundred-channel time-analyser for the same. When CIR became critical on 10 July 1960, a diffractometer was already in position. It took somewhat more than a year for the reactor to operate at a usuable power for scattering experiments. However, the centre of action was beginning to shift to CIR. By the end of 1961 several spectrometers had been installed there. Initially the Curran machine was used as a triple-axis spectrometer but was converted back to a diffractro-meter for magnetic scattering experiments as soon as a regular triple-axis spectrometer (Iyengar *et al* 1963) (figure 4.14) could be installed. The increase in count rates compared to APSARA was a great relief and a source of inspiration. Simultaneously, a beryllium detector spectrometer was also installed, to be followed soon after by a rotating crystal spectro-meter (Venkataraman *et al* 1963). The rotating crystal spectrometer had its own share of background problems partly because it was installed on a large twelve-inch beam hole. Consequently, it had to be redesigned for a four-inch beam tube. The triple-axis spectrometer was used first to measure the dispersion relations in magnesium and the other two spectrometers for measuring scattering from ammonium halides.

Of the group working in neutron scattering, Venkataraman, Usha Umakantha and myself had their first neutron scattering experience at Chalk River and Harwell. Later groups starting from Satya Murthy, K R Rao and Dasannacharya in 1958, were all initially trained at Trombay and then typically spent a year or a little more in some of the laboratories outside.

Figure 4.14 A triple-axis spectrometer at the CIRUS reactor, 1961.

Satya Murthy who had spent more than a year at Argonne National Laboratory returned in 1962. He continued further the magnetic scattering work. He initiated and built the polarised neutron spectrometer. A strong group developed in this area. Chidambaram had joined the group in 1962 and started chemical crystallography work on a new spectrometer which was soon updated to an automatic two-circle diffractometer (Chidambaram 1964). The multi-arm triple-axis spectrometer was also commissioned during this time (figure 4.15). By the end of 1964 the neutron scattering programme was well established at Trombay; experiments had been conducted on crystallography, magnetic structure, paramagnetic scattering, inelastic scattering from solids and quasi-elastic and inelastic scattering from liquids. The International Atomic Energy Agency held its third conference on 'Inelastic Scattering of Neutrons' in December 1964 at Bombay; this may be taken as an important landmark in the development of neutron scattering in India.

A final word on the pattern of development is relevant here. The growth

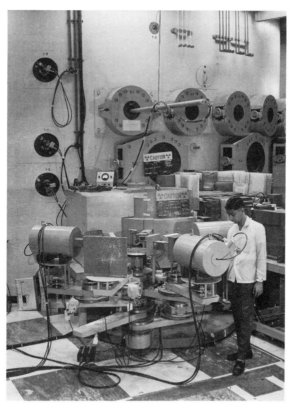

Figure 4.15 The multi-arm triple-axis spectrometer at the CIRUS reactor, 1964.

of this field was possible in India since we decided early that the development of neutron diffraction can take place only with local development of neutron spectrometers. The purchase of an imported spectrometer in 1958 would cost about an order of magnitude more than making one. Even if the engineering quality may have been somewhat less sophisticated nothing was sacrificed from the point of view of accuracy. Again, innovations in techniques grow when scientists think hard to overcome limitations on account of neutron flux. Some of the innovations in techniques like the use of guide tubes, cold moderators etc have come from groups working with smaller research reactors. A similar innovation I had made was to evolve the window-filter spectrometer making use of the cut-off from Be and BeO to improve the resolution compared to the usual inverted filter spectrometer (Iyengar 1964). It is strange that after nearly two decades we have built the ΔT spectrometer (Dasannacharya *et al* 1985) making use of the difference in the cut-off wavelength at two different temperatures to obtain the best resolution attempted so far. Such a spectrometer installed with the

Figure 4.16 P K Iyengar and M K Wilkinson at the Gatlinburg conference, June 1976.

SNS neutron facility at the Rutherford Laboratory has started yielding results.

It was thus the tremendous zeal generated among physicists in the early years of the atomic energy programme by scientists like H J Bhabha, K S Krishnan and R Ramanna which made it possible to install several spectrometers at CIRUS within four years and build a critical size group of scientists. Over the years this group has not only maintained the pace in neutron scattering work but also built complimentary techniques and devices to enhance the versatility of the research work.

References

Chidambaram R 1964 *Nucl. Physics Solid State Phys. (India)* **6B** 512

Dasannacharya B A, Goyal P S, Iyengar P K, Satya Murthy N S, Soni J N and Thaper C L 1985 *Proc. IAEA Conference Neutron Scattering in the Nineties*

Duggal V P 1961 *Nucl. Sci. Eng.* **6** 17

Duggal V P, Rao K R, Thaper C L and Singh V 1961 *Proc. Ind. Acad. Sci.* **53A** 59

Iyengar P K 1964 *Nucl. Instrum. Methods* **25** 367

Iyengar P K, Satya Murthy N S and Dasannacharya B A 1961 *Inelastic Scattering of Neutrons in Solids and Liquids* (Vienna: IAEA) pp 555–60

Iyengar P K, Venkataraman G, Rao K R, Vijayaraghavan P R and Roy A P 1963 *Inelastic Scattering of Neutrons in Solids and Liquids* vol. II (Vienna: IAEA) pp 99–110

Ramanna R, Sarma N, Somanathan C S, Usha K and Venkataraman G 1958 *Proc. 2nd U.N. Int. Conf. Peaceful Uses of Atomic Energy* A/CONF. 15/P/1636

Venkataraman G, Usha K, Iyengar P K, Vijayaraghavan P R and Roy A P 1963 *Inelastic Scattering of Neutrons in Solids and Liquids* vol. II (Vienna: IAEA) pp 253–63

4.7 The Beginning of Neutron Diffraction in Poland—Personal Reminiscences

J Leciejewicz

Institute of Nuclear Chemistry and Technology, Warsaw, Poland

The early history of neutron diffraction in Poland is closely linked to the rapid development of nuclear studies in Poland which took place after the decision, in the autumn of 1955, to purchase a research reactor from the Soviet Union. A group of scientists and engineers from the newly founded Institute of Nuclear Research, INR, soon paid a few weeks' visit to Moscow in order to identify what experiments could be started at this reactor. Among the visitors was Bronislaw Buras, deputy director of INR and solid-state physicist. In the early months of 1956 he organised a group of young scientists and technicians to prepare neutron-scattering experiments at the new reactor. This group included, among others, Denis O'Connor, Konrad Blinowski, Edward Maliszewski and Antoni Modrzejewski. After joining the INR in December 1955 I was invited by B Buras to set up a programme in crystal-structure studies because of my previous interest in X-ray crystallography. The reactor was built at Swierk, 33 kilometres south-east of the centre of Warsaw.

The time between 1956 and May 1958 was very exciting and busy for all of us: intensive reading of Bacon's *Neutron Diffraction*, frequent seminars, discussions, constructing equipment. Nobody had seen a neutron spectrometer with his own eyes and literature was rather difficult to get hold of. However, there was a lot of enthusiasm for new and exciting experiments, about which we heard from our few visitors, in particular Donald Hughes who gave a couple of lectures on neutron physics in 1957. Our first item of equipment was the control and data-recording system designed and built by Denis O'Connor, Konrad Blinowski and a group of technicians. Meanwhile large single crystals of metals (copper, zinc, lead and aluminium) for monochromators were grown successfully by Antoni Modrzejewski and Roman Czarnecki. The mechanical part of the spectrometer was designed by Denis O'Connor and Lucjan Bonkowski and constructed in the Machine Tool Factory in Raciborz to be ready in the spring of 1958, just before the reactor EWA (Experimental Water Atomic, Eva, the 'mother' of our reactors) was raised to its full power of 2 MW on

30 May 1958. All our reactors and critical assemblies were given female names e.g. EWA, MARIA, MARYLA.

The Solid State Physics Department of INR directed by B Buras had been transferred from Warsaw to Swierk one month earlier. All of us gave maximum effort to be ready in time. The reactor was officially started on 3 June 1958 by the Prime Minister, Mr J Cyrankiewicz, and there was a large celebration. Our own satisfaction was really high, since the only instrument ready for operation at the horizontal beam-holes was our neutron spectrometer at beam-hole No 4. The photograph taken just a few days before this event is shown below (figure 4.17). One can recognise at once how optimistic and inexperienced we were at that time concerning the instrumental background. In fact the monochromator shielding had to be strengthened very quickly. We used BF_3 counters produced at INR. Other equipment was made by ourselves, thanks to a well equipped mechanical workshop and a crystal growth laboratory available in Swierk.

The first experiment performed concerned the determination of the thermal neutron spectrum of the reactor. It was performed by a young graduate student, Jerzy Sosnowski, under the supervision of Denis O'Connor. In the course of this study the Renninger effect was 'rediscovered' and interpreted as 'parasitic' scattering (O'Connor and Sosnowski 1961). However, its practical implications for efficient neutron monochromatisation very soon became understood.

In late 1958 and early 1959 new experiments in neutron scattering were set up at EWA. A second neutron spectrometer was put at beam-hole No 2 (figure 4.18). This instrument was designed by Konrad Blinowski and a group of engineers as a double-axis unit, but in the course of time it proved to be most useful as a powder diffractometer. On this instrument neutron

Figure 4.17 The first neutron spectrometer constructed in Poland, at the reactor EWA Swierk in 1958.

diffraction studies of the crystal structure of the yellow modification of lead monoxide and tungsten monocarbide were completed in the summer of 1959, followed by red lead monoxide and tellurium dioxide. It is worth mentioning that the spectrometer at beam-hole No 2, later transferred to the beam-hole No 4, although very simple in construction, served us faithfully until 1969 and a large number of magnetic structure studies were performed with it. Furthermore, eleven spectrometers based on this instrument were built commercially between 1964 and 1968 at INR and sold to several nuclear centres in Europe and Africa.

During 1959–60 a cold-neutron facility was built by Denis O'Connor and Edward Maliszewski at the thermal column of the reactor. A beryllium filter cooled by liquid nitrogen and a chopper were used. This unit was intended for inelastic scattering studies, but turned out to be not very useful and the work in this field was done with a crystal spectrometer. Phonon spectra in metals were determined using single-crystal samples grown by Stanislaw Bednarski and Antoni Modrzejewski. The Inelastic Scattering Group of Jerzy Sosnowski, Edward Maliszewski and Stanislaw Bednarski has been active in this field until the present time.

At a very early stage Jerzy Janik and his collaborators from Krakow appeared in Swierk. This was perhaps in the late autumn of 1958 when I met for the first time Jerzy and Janina Janik, Szczesny Krasnicki, Franek Maniawski, Andrzej Murasik, Adam Wanic and perhaps a few others. They used an experimental set-up placed at beam-hole No 5. The transmission of slow neutrons in p-azoxyanisol in the nematic state was measured as a function of neutron energy and the applied external magnetic field. Scattering cross sections for hydrogen-containing molecules which exhibit

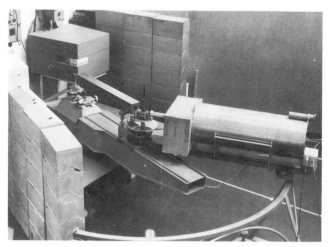

Figure 4.18 The second Polish spectrometer at Swierk, 1959, which served as a prototype for eleven instruments sold commercially.

vibrational energy levels in the thermal-neutron energy region were also studied. The first results of the above experiments were reported by Janik *et al* (1960).

Andrzej Murasik left the Krakow group about 1960 and joined in diffraction experiments at INR. His scientific interest was concentrated at that time on magnetic excitations in an iron–manganese ferrite of spinel type. A fairly large single crystal of a mineral, franklinite, became available. These studies were performed jointly with Adam Wanic and gave the spin-wave spectrum and structural information for this crystal (Murasik *et al* 1961).

The 'neutron club' at Swierk expanded rapidly during 1960–2. A graduate student from Warsaw University, Izabela Padlo, completed her MSc thesis on the refinement of the crystal structure of PbO_2. She was attached to the newly established Department of Nuclear Methods in Solid State Physics at Warsaw University, whose head was Professor B Buras. In 1960 Kazimierz Mikke came from the Reactor Physics Department of INR and started experiments on inelastic neutron scattering in ammonium halides; Andrzej Oles came from the Mining Academy in Krakow and began his programme on magnetic structures; Roman Ciszewski from Warsaw Technical University used our powder diffractometer for measurements on Mn_5Ge_3 and Mn_2Pd_3.

The period 1961–3 was also remarkable for intensive training abroad. In 1961 Konrad Blinowski went to Risø for two years to study critical scattering in chromium. I myself spent an unforgettable 13 months at Kjeller working with Arne Andresen on magnetic structures, and at the end of 1962 Andrzej Murasik went to Grenoble to study magnetic ordering in spinel-type oxides.

During 1962 Professor Buras initiated preparations for an experiment which was intended to demonstrate the usefulness of the time-of-flight technique for diffraction studies. The pulsed reactor at the Joint Institute of Nuclear Research at Dubna in the USSR was the obvious choice for such an experiment. Consequently, Jerzy and Izabela Sosnowska† went to Dubna at the end of 1962 where, together with the local people, they built a time-of-flight spectrometer. After my return home from Kjeller in December 1962, I was asked by Professor Buras to join the time-of-flight experiment in Swierk. First measurements were performed during February–March 1963. We used a simple Fermi chopper at the EWA reactor and a fairly large sample of silicon powder. The results (Buras and Leciejewicz 1963) turned out to be very inspiring.

To my mind, the beginning of 1963 marks the end of the pioneering era of neutron diffraction at Swierk. All groups which were active in Swierk

† Izabela Padlo married Jerzy Sosnowski in 1962: they both appear in the photograph (figure 4.10) taken at Dubna.

until the beginning of the 1980s crystallised at that time. It is true that one or two new experiments were set up later, some new people came and a new generation of spectrometers appeared in the 1970s, but the general framework of future activity existed at the beginning of 1963.

References

Buras B and Leciejewicz J 1963 *Nukleonika* **8** 78
Janik J A *et al* 1960 *Nukleonika* **5** 495
Murasik A, Ruta-Wala H and Wanic A 1961 *Physica* **27** 883
O'Connor D and Sosnowski J 1961 *Acta Crystallogr.* **14** 292

4.8 Neutron Diffraction in Australia

T M Sabine

*New South Wales Institute of Technology,
Sydney, Australia*

Australia in the 1950s was a prosperous and optimistic country. Its economy was boosted by the increase in birthrate following World War II and by the intake of migrants, predominantly from Britain. Primary industry, upon which it relied almost entirely, was booming with a dramatic increase in wool prices following the war in Korea.

Australia was developing a technological economy. The universities commenced to award PhD degrees in 1949. The CSIRO was formed in the same year with the objects of scientific research and application of results, the maintenance of measurement standards, and the publication and dissemination of scientific information.

Manufacturing industry was developing behind a protective wall of tariffs which, it was expected, would be dropped as the industries grew to maturity.

The year 1955 was the high tide of euphoria concerning the peaceful benefits of atomic energy, with its promise of 'electricity too cheap to meter', culminating in the first Geneva conference.

Australia was anxious to obtain the benefits of the new 'perennial fountain'† and, after passing the Atomic Energy Act in 1956, the Government of Australia formed the Australian Atomic Energy Commission and turned to England for help in establishing a research programme. This seemed perfectly natural at the time. While the Australian continent is physically on the edge of Asia, the country behaved and thought as if it was moored in the Irish sea. Britain was the 'mother country' and 'home'. Oddly this love of England was increased by the extraordinary cost of travel to and from Australia. These barriers to travel were kept up by the Anglophile Prime Minister R G Menzies, whose successive governments ruled from 1949–72.

Australia was originally a British colony acting as a source of food and raw materials for England. Much of the colonial dependence has lingered on to the present day.

Uranium mined in Australia in the early 1950s provided fuel for Britain's ambitious nuclear programme. Between 1952 and 1956 R G Menzies permitted extensive tests of British atomic weapons first on the island of Monte Bello off the west coast and later in the central Australian desert at Maralinga. It was only in 1985, through a Royal Commission set up by the Australian government, that information began to be reluctantly divulged on the fallout pattern from these tests.

As a consequence of this special relationship the scientists who were to build the AAEC were posted *en bloc* to Harwell.

At that time I was a graduate student in the Baillieu Laboratory of the University of Melbourne with the fortune to be under the supervision of J N Greenwood of the University and R C Gifkins of CSIRO. After a short period in the CSIRO Division of Tribophysics under W Boas, I joined the AAEC and was given the task of developing neutron diffraction. The Chief Scientist, C N Watson-Munro, had clearly seen the importance of this technique for solid-state studies. I shall never forget the luxury of that 1956 first-class flight to England. The cost of the tickets for my wife and myself was equal to my annual salary.

At Harwell I joined the Metallurgy Section of AAEC under K F Alder and was placed in G E Bacon's neutron diffraction group. Others in the group were N A Curry, W E Gardner, R D Lowde, R F Dyer, D A Wheeler and B T M Willis.

The time at Harwell was enjoyed by all members of the Australian contingent. Under the employment conditions operating at the time we

† The revelation of the secrets of nature, long mercifully withheld from man, should arouse the most solemn reflections in the mind and consciences of every human being capable of comprehension. We must indeed pray that these awful agencies will be made to conduce peace among nations, and that instead of wreaking measureless havoc upon the entire globe, they may become a perennial fountain of world prosperity.

Winston Churchill, 6 August 1945

were paid a generous hardship allowance for living so far from home. Several episodes stand out: on the scientific side a lecture by A H Cottrell at the Royal Society; on the social side a cricket match against the English Metallurgy Division, won by the English (we maintain) through the 'importation' of the Oxfordshire fast bowler; an Australian Metallurgy Section tour of the UK nuclear industry establishments in the north of England. This ended in a boat race on Lake Windermere. The stony faces of the other guests on our return to the hotel reminded us that we had forgotten how well sound travels over water.

After being introduced to Beevers–Lipson strips by Bill Gardner and the Wantage pub by Terry Willis I returned to Australia in 1958 through the USA, meeting C G Shull, E O Wollan, M K Wilkinson, J Hastings, L M Corliss and W C Hamilton.

The site for the Australian nuclear laboratory had been chosen to be Lucas Heights, 40 km south of Sydney. Construction commenced in 1955, and HIFAR (High Flux Australian Reactor), a copy of the British DIDO, went critical on Australia Day (January 26), 1958. Full power operation (10 MW) was achieved in 1960.

Lucas Heights was situated in the Shire of Sutherland, which was a rapidly growing suburb, but was still largely unsewered. The depot to which the raw sewage was brought by truck was next to Lucas Heights. The smell is a vivid memory for all those who worked there.

Into this environment the Australian contingent from Harwell was dumped to cope with rapacious real estate salesmen and loss of the hardship allowances that had been paid for living in places such as Oxford.

The main thrust of the research programme was to be the development of a high-temperature gas-cooled reactor system based on beryllium oxide. This concept, which was due to C G J Dalton, Chief Engineer, was thought most likely to produce a reactor suitable for the Australian environment, and to provide a vehicle for the training of nuclear technologists. From the beginning it was agreed that a small fraction of the Commission's resources would be devoted to pure research. This included neutron diffraction which was the responsibility of the Materials Division led by R Smith.

Initially two neutron diffractometers were to be installed on holes 4H1 and 4H2. The collimator design was copied from the designs already prepared for DIDO. They were constructed in Australia and inserted into HIFAR in 1961. The 4H2 collimator was a tight fit and, in a spirit of elan characteristic of the times, C A Logan and I used a hydraulic jack to finish the job. In 1985 the removal of this collimator was an engineering problem of some magnitude.

A copy of the DIDO single-crystal diffractometer and an Australian designed and built powder diffractometer were installed by 1961. Following a suggestion by B S Hickman, a long-wavelength spectrometer to study defects in irradiated BeO was designed and built. This used an Australian mica crystal with Be and BeO filters fabricated in the ceramics section by K D Reeve.

The first results were published by Sabine, Pryor and Hickman (1961). The work on BeO was finished by 1964. Its achievement was the demonstration at a fundamental level that BeO was entirely unsuitable as a solid moderator in a nuclear power reactor.

In 1958 a decision was made by government to double the number of universities and to support a significant increase in university research work. Many universities took the view that training in nuclear science and engineering would be essential for the future Australia. They therefore looked to the acquisition of reactors on each campus.

J P Baxter (Chairman of AAEC) and Prime Minister R G Menzies foresaw the dangers of this approach and Cabinet authorised the establishment of a joint organisation between the universities and AAEC. All the universities agreed to join on the basis that the AAEC would match their funds one for one. At the inaugural meeting on 4 December 1958, Baxter said

> Not even the wealthiest university can equip itself completely for the study of nuclear energy today. The universities provide basic training in nuclear science and engineering with their own equipment and staff, but in any advanced programme they must have access to a nuclear reactor and specialised chemical, engineering and metallurgical laboratories to handle highly radioactive materials.

The joint organisation was called the Australian Institute of Nuclear Science and Engineering (AINS&E). It was provided with a building 'outside the gate' at Lucas Heights. E A Palmer was appointed Executive Officer in early 1960. From the beginning AINS&E requested applications for research grants from members, had these assessed by expert committees and, if the assessment was positive, provided credits for Lucas Heights facilities and for travel. All projects were treated equally irrespective of actual travel costs. This method was also adopted later at the Institute Laue–Langevin in Grenoble.

Neutron diffraction was of immediate interest to the universities and the then President of AINS&E, D O Jordan, called a two-day seminar in November 1959.

At that time crystallography was very much more advanced in CSIRO than in the universities. J M Cowley and A F Moodie in electron diffraction, A McL Mathieson in organic X-ray crystallography, B Dawson in theoretical crystallography and A D Wadsley in inorganic structure analysis had established a strong Australian commitment to the field.

In the universities, G H Cheeseman had an active group at Tasmania, H C Freeman had returned to Sydney from CalTech and E N Maslen to Perth from Oxford. E O Hall had joined the University of New South

Wales, and R Street had taken up the Foundation Chair of Physics at Monash University. These appointments, and the interests of the appointees, biased effort toward single-crystal diffractometry for structure analysis and powder diffractometry for magnetic structure studies and studies of alloys.

With the active support of C J B Clews, who held the Chair of Physics at the University of Western Australia, the first structure study was completed and published in 1961 (Clews, Maslen, Rietveld and Sabine 1961). H M Rietveld, who was a graduate student at the time, later became famous for his invention of the Rietveld method.

Development of single-crystal and powder studies proceeded through the 1960s. There was a close working relationship between the reactor operating staff and the neutron beam experimentalists, and the around-the-clock help of the rig technicians was of great assistance. It enabled the introduction of automatic three-dimensional data collection for single-crystal problems in 1963. I had generalised the angle setting formulae for the General Electric Goniostat to crystals of any symmetry. The angle settings were written on cards supplied to the reactor staff. They set these every twenty minutes and started a paper tape punch data recording system. These tapes were processed later to yield integrated intensities using systems written by Suzanne Hogg.

B M Craven, who came to Sydney for an extended stay in 1964, worked with G W Cox to get a set of crystallographic programmes operating on the AAEC computer. These programmes, the availability of reactor staff and the Commission decision in 1962 not to charge AINS&E for neutrons, led to a considerable programme of single-crystal studies. F H Moore was appointed to AINS&E to supervise this programme in 1967.

D A Wheeler had joined the AINS&E neutron diffraction group from Harwell in 1965. His major task was the provision of liquid helium facilities for powder work. This diffractometer served the needs of a generation of PhD students from Monash University. Among them was T J Hicks, who later took over a mechanical monochromator constructed for studies of defects, and built Longpol, with which a programme of diffuse scattering from magnetic systems was initiated by him and his students.

In 1964 the AAEC made a decision to build a triple-axis spectrometer. Under the guidance of A W Pryor, this was designed at Lucas Heights and built in Australia by Vickers-Ruwolt of Melbourne. In 1967 Margaret Elcombe arrived from Cambridge to take responsibility for commissioning the instrument and introducing Australian solid-state scientists to lattice dynamics.

Other instruments were installed through the 1960s. In particular a collimator which viewed the whole source was installed on a six-inch hole. The very high flux of the monochromatic beam was used to carry out structure analysis with very small crystals. This development led to my work with

E V Weinstock at Brookhaven which showed that maximum flux at the specimen was provided by viewing the whole source. This philosophy was adopted by the ILL.

Later this hole was used for the high-resolution powder diffractometer, the first such instrument on a conventional source to give a performance equal to that of D1A at the ILL.

A G Klein of the University of Melbourne initiated fundamental studies in neutron optics. However, because of the lack of a cold source these were transferred to the ILL. Over the years strenuous efforts have been made to obtain such a source but national scientific priorities have not allowed it.

Neutron diffraction has made a strong impact on post-graduate education for research and development. It is a field which offers academic rigour, advanced technology, and experience in sharing the use of a major facility.

The first PhD degree involving the use of neutron beams was awarded in 1961. By 1982 89 PhD degrees and 10 Masters degrees had been awarded. Among those graduates are H M Rietveld (1961), J C Taylor (1964), T J Hicks (1965), E R Vance (1968), A W Hewat (1970), S A Mason (1971), Sylvia Mair (1974), J E Tiballs (1974), Barbara Moss (1979) and J Parise (1980). These scientists would have been invaluable in a national development plan for science and technology. However, and tragically, Australia has not pursued such a plan. Many of the graduates went to overseas laboratories and stayed. The favourite overseas laboratory was Harwell, both because of the earlier connection and because of the collaboration of B Dawson, B T M Willis and A W Pryor. Australian scientists formed the nucleus of the British group at the ILL.

Australia has been active in the International Atomic Energy Agency programme for the effective use of research reactors in South-east Asia. These reactors had been provided under President Eisenhower's *Atoms for Peace* initiative in 1955. An example of this active collaboration was a paper published in *Acta Crystallographica* in 1967. The authors are: G W Cox, T M Sabine (Australia); V Padmanabhan (India); Nguyen Tu Ban (South Vietnam); M K Chung (Korea); A J Surjadi (Indonesia). Cooperation with Asian countries has continued through regional agreements arranged by IAEA and by bilateral science and technology agreements between Australia and neighbouring countries.

A considerable amount of thought and discussion is taking place about the way Australia should satisfy its requirements for neutrons into the 21st century. Whilst Australia, relative to its population of 16 000 000, has the same fraction of scientists as other OECD countries, the absolute size of its scientific community is too small to support a high-flux reactor or a spallation source. There will be a large scientific community in the South-east Asian region in the years ahead and Australia has been active in founding the South-east Asian Regional Crystallographic Association.

This association embraces those countries within the region bounded by Japan, China, Pakistan, India, Australia and New Zealand. One of the objects of the association is to foster the establishment within the region of research facilities to be shared by all members. The combined population of countries within the region is two billion. Realisation of this plan will take one or two generations. Until then Australia is looking towards a mechanism to enable its scientists to share in the facilities currently available in Europe and USA.

It is proposed to continue using HIFAR as an active centre for neutron research. There are many experiments which can be carried out on medium-flux reactors. It is also essential to have a substantial facility to which graduate students have easy access, and which can be used as a field trip or excursion for undergraduates. Active involvement of students in parts of the AAEC programmes is a beneficial effect of the establishment of AINS&E. In Australia the export of uranium is a matter for intense public debate. Experience at Lucas Heights provides non-specialists in atomic energy with information for the debate.

References

Clews C J B, Maslen E N, Rietveld H M and Sabine T M 1961 *Nature* **192** 154–5
Sabine T M, Pryor A W and Hickman B S 1961 *Nature* **191** 1385–6

4.9 Neutron Scattering in China

Z Yang

Institute of Atomic Energy, Beijing, China

A Brief History

Neutron diffraction work in China began in 1959, at the Institute of Atomic Energy, Peking where the first heavy-water research reactor HWRR, with a power of 10 MW, went critical in the autumn of 1958. The first research group of four people was then formed under Professors He and Dai, now the director of IAE. The first equipment constructed was a neutron spectrometer, with a primitive monochromator stage taken from an old X-ray diffractometer of the 1930s. This spectrometer was used both for diffraction and as a monochromatic neutron source. Another diffractometer of better design, with remote control and 2:1 motion, was installed at the end of 1959. These two instruments served as the basic equipment for diffraction study in China during the 1960s. Studies of both crystal structure and magnetic materials were carried out in that period; samples of SnO_2, $BaSO_4$, $ZrSiO_4$, $(Ni-Zn)Fe_2O_4$, $BaFe_{12}O_{19}$, MnAl etc were examined. The first observation of the strong extinction change brought about by piezoelectric excitation of a quartz crystal was also studied at that time as a new effect. However, because of various factors most of this work was unpublished or only published much later. Some references are given at the end of the article.

The period from 1966–70 was full of political movements in China and the diffraction work practically stopped. Recovery began slowly at the beginning of the 1970s with the reconstruction of the two old spectrometers, followed by the design of the first inelastic scattering equipment (TOF) in 1973 when Z Yang was appointed as the head of this group. Field-induced extinction change work was restarted and the anomalous effect in α-LiIO$_3$ was soon observed. The neutron diffraction group began a more systematic study of magnetic materials and rare-earth–transition metal compounds in cooperation with Peking University, together with the

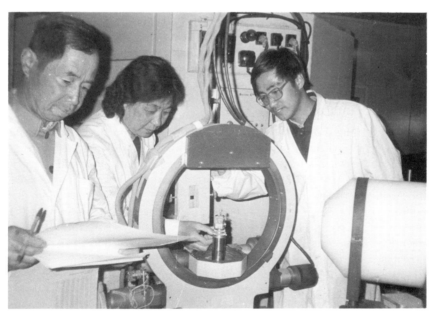

Figure 4.19 Z Yang, Y F Cheng and Q Li at the single-crystal diffractometer in the Institute of Atomic Energy, Beijing, 1985.

study of garnet-type ferrites and other materials such as Mn–Al–C and
Al–Fe–B. The old neutron spectrometer was modified into a Be-filter
spectrometer for inelastic work. Phonon properties of a series of metal
hydrides, as well as some chemical substances and superconducting materials, were studied, together with time-of-flight. The size of the group was
increased from 4 to 15.

Rapid development occurred between 1979 and 1982 when scientists
from the Institute of Physics headed by Professor Z Zhang, began collaboration with the Institute of Atomic Energy. At the same time international contacts developed. A new triple-axis, 4-circle, spectrometer was
constructed in cooperation with France and installed at the end of 1983.
Construction of an improved time-of-flight machine, SANS, with a cold
source, was commenced and it is expected to be in use in 1987.

International contacts and collaboration within China

In the early days researchers in neutron scattering were almost isolated
from the outside world except for a few chances to participate in conferences of the Joint Institute of Nuclear Research in Dubna. Cooperation
abroad began at the end of the 1970s when extensive contact with France
suggested the possibility of building an 'Orphee'-type research reactor. This
ambitious plan was dropped due to lack of funds but cooperation with
Saclay people started. At the same time, under the more friendly atmosphere for western countries, many scientists from developed countries
visited China, such as J Higgins from England, D Cribier from France, H
Stiller from FRG, Y Ishikawa from Japan, S H Chen and S Werner from
the USA and others. A number of Chinese scientists began to work in this
field abroad; these include people specially sent out from IAE for practice
in neutron scattering as well as those who got into this field by chance.
After returning home they began to look for opportunities of doing
neutron work in China, and because they worked in a variety of districts
and institutions, this further promoted cooperation within China. Propa-
ganda for this new technique began with a small conference organised by
the Academy of Science of China in the spring of 1979, in which about 200
people from various institutions participated. Then, in the autumn of the
same year, a comprehensive set of lectures was given by three French
scientists from Saclay. Another series of lectures followed in 1980, given by
German scientists as a part of a scientific exchange programme between
China and the Federal Republic of Germany. The first conference of
Chinese workers in neutron scattering was held in 1983 and this signals the
revival of Chinese neutron scattering work. At present there are 29 people
regularly working in the neutron scattering laboratory at IAE, headed
since 1984 by Dr C T Yeh, with 7 people from the Institute of Physics

headed by Dr Q Ling. There are at least 10 others who worked, or who are still working, abroad in this field and who belong to other institutions than these two.

The main permanent collaborations, now well established, are with Peking (Beijing) University, for the study of rare-earth–transition metal compounds; the North-East Technological Institute, for amorphous magnetic materials and other magnetic compounds; Nankai University for hydrogen-storage materials, including metal hydrides; and the Institute of Non-ferrous Metals, for superconducting materials. There is also short-term cooperation on some special topics including crystal-structure study, chemical properties and the study of catalysis.

References

An Wan-shou, Zhang Huan-qiao, Yang Ji-lian, Chu Chia-hsuan and Li Guang-ding 1961 *Acta Physica Sinica* **17** 222 (abstract in Russian)
Chu Chia-hsuan, Cheng Yu-fen, Shen Chin-chuan 1965 *Scientia Sinica* **XIV** 1541
Yang Zhen, Zhang Huan-qiao, Zhou You-po 1974 *Acta Physica Sinica* **23** 400
Zhang Huan-qiao 1963 *Acta Physica Sinica* **19** 477 (abstract in Russian)

4.10 Progress in Japan

S Hoshino

Institute for Solid State Physics, University of Tokyo, Japan

It was 1961 before neutron diffraction experiments could be made in Japan, one and a half decades behind the advanced countries in this field. Construction of atomic reactors in Japan was tremendously delayed because research in the field of atomic energy was prohibited for some time after World War II. In 1955 the Japan Atomic Energy Research Institute (JAERI) was founded and then research in atomic energy and related fields commenced in Japan, entirely restricted to peaceful applications.

Up to that time, solid state physicists as well as crystallographers in Japan had felt deeply chagrined at the critical situation which had arisen after fifteen years' delay in starting any experimental research using neutrons to study the structure of matter. As soon as JAERI was established construction work on a CP-5 type research reactor JRR-2 began. In 1957 a Neutron Diffraction Committee was organised in the Science Council of Japan as a joint subcommittee of the National Committee of Crystallography and that of Pure and Applied Physics, The Members of this Neutron Diffraction Committee were S Kaya, T Watanabe, T Hirone, T Nagamiya, Y Kakiuchi, T Fujiwara, R Sadanaga, S Miyake and T Tanaka. As a result of the intense efforts made by this committee, two beam holes of JRR-2 were allocated for neutron diffraction research, one for the use of the Division of Solid State Physics in JAERI and the other for the joint use of the Institute for Solid State Physics (ISSP) of the University of Tokyo and the Electrical Communication Laboratory (ECL) of the Nippon Telegraph and Telephone Public Corporation.

In 1958, N Kunitomi of JAERI visited Brookhaven National Laboratory for about three months to study magnetic substances by neutron diffraction. After returning from Brookhaven, he organised a special committee in JAERI and began to plan the construction of a powder neutron spectrometer. Subsequently, S Miyake of ISSP started to make a conceptual design for further spectrometers with K Suzuki of ECL and others. Afterwards in 1960, S Hoshino, who had spent six months in Brookhaven studying ferroelectric phase transitions by neutron diffraction, was appointed to take charge of neutron diffraction work in ISSP. A detailed plan was then produced to construct two spectrometers, one for ISSP with an energy analyser and the other for ECL with polarised neutrons.

JRR-2, the first high-power research reactor in Japan, with a thermal neutron flux of about 10^{14} neutrons cm^{-2} s^{-1} at full power (10 MW), went critical in 1960 and test operation under the power of 2–3 MW started in 1961. At the same time, the three neutron spectrometers mentioned above were completed (figures 4.20, 4.21) and the neutron diffraction measurements which had been a long-cherished desire by Japanese scientists now became feasible.

In September of 1961, just in time, the International Conference of Magnetism and Crystallography was held in Kyoto. At that conference, three papers concerning neutron diffraction were reported as the first work performed in the Japanese institutions. Two of them were about the instrumentation of the newly constructed neutron spectrometers and the other was a study of the structure of uranium-3d transition element alloys. In addition, some Japanese scientists reported their works carried out outside the country. Among them, Y Yamada of Osaka University returned from MIT and M Sakamoto of JAERI, from Chalk River, having finished their one-year studies of neutron scattering under the guidance of C G Shull and of B N Brockhouse, respectively. Among participants at the conference were many pioneer, as well as active, scientists in the field of neutron scattering and diffraction such as E O Wollan, C G Shull, G E Bacon, M K Wilkinson, R D Lowde, W C Koehler, J M Hastings, A D B

Woods, T Riste, H Dachs, B C Frazer, R P Ozerov, G Caglioti, T M Sabine, S W Peterson and G Shirane. Their visit to Japan in that year gave a valuable stimulus to Japanese scientists who had just started experimental work using neutrons.

Figure 4.20 Three early spectrometers at the JRR-2 reactor in 1961.

Figure 4.21 S Hoshino at the JRR-2 reactor in the early 1960s.

Thus, experimental research in neutron scattering and diffraction started in Japan. During the early years, however, the operation time of the reactor was so short that the available machine-time for neutron work was limited. Continuous operation of the reactor at 5 MW began in 1964, reaching the full power of 10 MW in 1965. Nevertheless, a regulation for operating atomic reactors in Japan was very restrictive and the annual rate of the reactor operation at its full power was confined below 40%. Actually it was about 30% or less. Therefore, development of neutron studies in Japan made little progress in the early stages. During these years, Y Ishikawa stayed at Grenoble, Y Hamaguchi at Oak Ridge, A Okazaki and K Hirakawa at Harwell, and so on; a number of Japanese scientists enriched their experience of neutron work with a stay at foreign institutions.

In 1965 JRR-3, with one tenth of the neutron flux of JRR-2, was constructed in JAERI. A double-axis neutron spectrometer for the joint use of Tohoku University, Osaka University and JAERI, and a chopper TOF spectrometer of JAERI were soon installed at this reactor. Further, the Research Reactor Institute of Kyoto University was founded in 1963, and KUR, a light-water moderated tank type reactor, achieved a nominal maximum power of 1 MW in 1964; the power was increased to 5 MW in 1967. Neutron diffraction equipment with four-circle motions designed by I Shibuya was installed at KUR. In this way, the total machine time available for neutron work increased gradually. Nevertheless, Japanese scientists still felt uneasy about their status, which was backward in view of the progress being made elsewhere in the world. Although JRR-2 with its thermal neutron flux of 10^{14} was a relatively high-flux research reactor when it started its operation, it was soon overtaken by the high-flux reactors which were constructed at Brookhaven and Oak Ridge in the USA in 1966 and a rapid development of neutron scattering research and techniques was expected. Moreover, the Institut Laue–Langevin was founded in 1967 and the construction of high-flux reactors was scheduled in Europe, too.

Drawing conclusions from these developments abroad, scientists in this field organised a neutron diffraction research group in 1966 with the support of a Grant-in-Aid from the Ministry of Education, Science and Culture. This research group consisted of about twenty scientists and went into action to achieve cooperative research on the development of techniques for neutron scattering experiments. The group also had extensive discussions on future projects that should be promoted in Japan and, in 1968, drew up a plan which consisted of the following three items.

(1) A high flux reactor with a thermal neutron flux of the order of 10^5 neutrons cm^{-2} s^{-1} should be constructed in Japan.

(2) A pulsed neutron source using a spallation nuclear reaction should be developed.

(3) More effective utilisation of the existing reactors should be considered and, at the same time, cooperative research with overseas countries should be proceeded with.

According to these fundamental policies an appeal was made to the Government as well as to other organisations such as the Science Council of Japan, JAERI and other institutions. Later, an executive committee was set up in the research group to promote these proposals. Members of the committee were N Kunitomi, S Hoshino, K Hirakawa, Y Ishikawa, Y Hamaguchi, I Shibuya, S Komura, J Harada and Y Nakamura.

As for the first project, the Science Council of Japan recommended the construction of a high-flux reactor to the Government in 1971 and it was decided that it should be built at Kyoto University and would have a thermal power of 30 MW, in order to provide greater opportunities for neutron beam experiments and neutron irradiations, including medical applications. A detailed design for KUHFR was made by the Research Reactor Institute and the budget was approved by the Government in 1975. However, the reactor has not yet been realised, at the time this article is being written, because of an environmental problem at the Research Reactor Institute located in the southern Osaka prefecture. Namely, people who live in the vicinity of the Institute have not agreed to the construction of the KUHFR. Japan is the only country which has ever been attacked by atomic bombs and public opinion concerning nuclear energy is exceptionally critical and sensitive. Therefore, the construction of the high-flux reactor may have to be modified to some extent.

On the other hand, the second project, for the development of the pulsed neutron technique, has progressed steadily. As soon as a 300 MeV electron linac was completed in Tohoku University in 1967, M Kimura and his colleagues started neutron diffraction measurements, using a spallation neutron source with a lead target surrounded by light water. It was found that the diffraction patterns obtained for powders and single crystals of some metals could be compared favourably with the data taken at the pulsed reactor at Dubna by B Buras, or with those taken by the chopper time-of-flight method at steady-state reactors. Since then, further developments of neutron diffraction and scattering methods with the pulsed neutron source have proceeded extensively. Such development studies with the electron linac in Tohoku University prepared the ground for the later establishment of neutron scattering facilities by utilising a proton synchrotron. It can be said that the work with the spallation neutron source carried out in Japan led the world. This is certainly due to the farsightedness of M Kimura and to the enthusiastic efforts made by Y Ishikawa and his colleagues.

Progress has also been made on the third project of the future plans, i.e. the effective utilisation of the existing reactors and international cooperation. During the years following 1968, various kinds of neutron spectrometer were installed at JRR-2 and JRR-3 by ISSP and Tohoku University, and the older instruments have been progressively improved or remodelled. Technical improvements, especially the introduction of pyrolytic graphite monochromators, enabled us to measure inelastic neutron scattering more easily. Thus, the quality as well as the quantity of the research in the field of neutron scattering and diffraction progressed step by step in those years. In the early 1970s, a number of Japanese scientists carried out experiments with USA scientists using HFBR and HFIR under the USA–Japan cooperative research programmes.

So far I have described objectively the progress of neutron scattering and diffraction in Japan up to the early 1970s. As requested by the Editor, I have explained 'how we started, how we were trained and what the early difficulties and frustrations were'. Therefore, the detailed content and the nature of our studies have scarcely been mentioned: they are well reported in the literature. It can be said that progress in the early stages in Japan was by no means rapid but, as a whole, it was made steadily by overcoming various difficulties. As for the recent progress, a brief account is given below.

In the past decade, many more developments have taken place in Japan. As a successor to the pulsed neutrons provided by the electron linac in Tohoku University, we established KENS in 1980 as a national centre for neutron scattering studies, by utilising a booster ring on the high energy proton accelerator at the National Laboratory for High Energy Physics, KEK. There, we now have a cold-neutron source and various types of spectrometer, and a great number of scientists in many fields are using these facilities in addition to those at reactors. At the same time, at the reactors, new spectrometers including Tanzboden-type ones, were installed at JRR-2, and at KUR a guide tube has been provided to measure small-angle scattering and a cold source installed very recently. Besides these changes, a long-term USA–Japan collaboration project started in 1981 under the Agreement on Cooperation in Research and Development in Science and Technology between the Governments of USA and Japan. Recently, a Japanese-funded advanced polarised-neutron spectrometer and a wide-angle spectrometer have been installed at Brookhaven and Oak Ridge, respectively. Equipment for making studies under extreme conditions such as low temperature and high pressure has also been installed at Oak Ridge. Though we have met difficulty in constructing our high-flux reactor, JAERI started construction work on a remodelled JRR-3 with a cold source and guide tubes as the successor to the JRR-2 research reactor. A future plan to construct a more intense spallation neutron source is at present under discussion.

In concluding this article, I would like to refer to words by T Nagamiya in his welcoming address at the opening of the Sixth Yamada Conference on Neutron Scattering of Condensed Matter, held at Hakone in 1982: 'Japan was once lagging behind other countries, but now I believe we have reached a high international standard.'

Acknowledgment

The author would like to thank Professors N Kunitomi, K Suzuki, I Shibuya, Y Ishikawa† and N Niimura for their cooperation in collecting the information for writing this article. The author's thanks are also due to Professor B Buras who kindly sent the first draft of his article on 'The time-of-flight diffraction method'.

† *Note added in proof:* Y Ishikawa died suddenly in February 1986. His death represents a great loss to the neutron scattering community in Japan.

5 The Changing Scene: The Appearance of the High-flux Reactor

5.1 High Flux at Brookhaven

L Passell

Brookhaven National Laboratory, USA

Late in 1954, Donald J Hughes and the group of young scientists who worked with him found themselves face-to-face with a discouraging problem: their recently completed fast chopper time-of-flight system was not fully resolving the resonance structure of the epithermal neutron cross sections they were trying to measure. And—to make matters worse—lengthening the chopper flight path, the usual way to improve resolution, was ruled out by the low flux of the Brookhaven Graphite Research Reactor (BGRR): the intensity loss would have been prohibitive. Trying to salvage what he could of one of his major research programmes, Hughes arranged a collaboration with the fast chopper group at Canada's Chalk River Laboratory where the high-flux, NRU reactor was soon to come on-line. But he lost no time in pointing out to Leland J Hayworth, Brookhaven's Director, that the day was not far off when the Laboratory would need a higher flux reactor to maintain its position in the neutron field.

Hayworth had long suspected as much and asked Hughes to organise a Reactor Study Group and begin making plans for a high-flux replacement for the BGRR. The group that Hughes assembled looked first at the over-all requirements of the Laboratory's experimental programme. What were primarily needed, they decided, were external beams; demand for facilities for either isotope production or in-pile engineering studies was clearly going to be limited. Next, the possibility of building two reactors was discussed: one to serve as a thermal neutron source and another for epithermal neutrons. This was dismissed out of hand as too costly. Finally, they turned to the most difficult question of all: finding a reactor configuration that would give substantially increased external neutron beam intensities.

Jack Chernick, the Laboratory's reactor physics specialist, was given the task of exploring the available options. Eventually he narrowed the choice down to a small, undermoderated core arrangement with either a heavy-water, graphite, beryllium or beryllium oxide reflector and with either heavy-water, liquid sodium or (possibly) a liquid sodium–potassium eutectic mixture as coolant. Further analysis led him ultimately to decide on a heavy-water moderated and cooled core surrounded by a heavy-water reflector. He estimated that a reactor of this type would produce a peak thermal neutron flux of about 5×10^{14} neutrons cm^{-2} s^{-1}. Moreover, the high flux could easily be translated into high external beam intensities because the thermal flux peaked, not in the core, but in the reflector where the beam tubes would be located. Julius M Hastings, another member of the Study Group, proposed a further refinement: mounting the beam tubes tangentially rather than radially. This, he pointed out, would decrease fast neutron contamination without materially altering the thermal beam intensities.

Chernick's design was adopted in 1956 and after some further discussion a decision was made to install nine beam tubes, eight tangential and one radial, to serve as an epithermal source. Soon afterwards a contract was let to the General Nuclear Engineering Corporation for a detailed heat-transfer study, heat transfer being viewed as the crucial factor in determining the ultimate performance of the reactor. In July 1957, with a favourable report from General Nuclear in hand, the Laboratory submitted a proposal to the Atomic Energy Commission (AEC), its parent federal agency, to build a high-flux beam reactor (HFBR).

The AEC's initial response was not encouraging. Having been on an austerity budget for two years past, it saw no real possibility of obtaining the funds needed for a project on the scale of the HFBR. And to make matters worse, Brookhaven's sister laboratories, Oak Ridge and Argonne, not to be outdone, had also submitted proposals to build new research reactors. Without funds for even one such facility, the AEC suddenly found itself being called upon to support the construction of three!

What saved the day was something totally unexpected: the launching of Sputnik, the world's first orbiting satellite, by the Soviet Union in the autumn of 1957. The Soviet challenge to American technological leadership produced an immediate public reaction. Pressure was put on the Government to expand support for both science and technology and in March 1958 Congress amended Dwight D Eisenhower's Presidential budget to include extra money specifically for research. Realising that it had a golden opportunity, the AEC quickly decided to proceed with Brookhaven's HFBR and with Oak Ridge's high-flux irradiation reactor (HFIR). Argonne's 'Mighty Mouse' project, which was far more ambitious, was put aside. To the great surprise and joy of the Brookhaven staff, the Laboratory was suddenly asked to submit a formal proposal to be included in the 1960 fiscal year federal budget. Dated July 1958, and designated USAEC project 562-60, the document estimated the cost of the reactor at ten million dollars and two years as the time required for its construction.

Chernick by this time had made considerable progress in defining the HFBR core and heavy-water moderator and reflector configurations and, with the help of General Nuclear, had worked out the arrangement of its primary cooling system. His design is described in essentially its final form in a paper he and his co-workers presented at the 1958 Geneva Conference on the Peaceful Uses of Atomic Energy (Auerback *et al* 1958).

With funding assured, Herbert J C Kouts, the head of Brookhaven's experimental reactor group, set to work building a zero-power mock-up of Chernick's design. He and Joseph M Hendrie, the project engineer, soon discovered, to their dismay, that General Nuclear had considerably underestimated the power needed to reach the expected level of performance. To everyone's relief, however, ways were eventually found to get around the difficulties and ultimately it proved possible to achieve a peak flux of about 1.6×10^{15} neutrons cm^{-2} s^{-1} with only a modest increase in power. Most of 1959 was spent finalising the core size and shape and determining the fuel-element characteristics, control capability, flux distributions etc, as well as in finding the optimum arrangement of beam tubes. By 1960 work had progressed to the point where outside firms could be engaged to prepare detailed construction drawings. A year was spent on site reports and final cost estimates. Finally, in the spring of 1961, the last required document, a preliminary hazards report, was forwarded to the AEC. Approval to begin construction arrived at the end of the summer. Site clearing took a few weeks, and then, in early autumn, work on the reactor began.

In October 1965, almost four years to the day after the first shovelful of earth was moved, the HFBR stood completed and ready for testing. It had taken twice as long to build as projected but cost only 12 per cent more than the 10 million dollars originally estimated. Sadly, Donald Hughes, the man who, with Chernick, had been most responsible for its creation, never lived to see construction begin. He died, quite suddenly, at 45 years of age, in April 1961, just as the last of the safeguard reports was being forwarded to the AEC. Chernick was more fortunate. He survived Hughes by ten years and had the satisfaction of seeing the HFBR completed and in routine operation before his death at age 60 in 1971. They are both shown in figure 5.1.

Figure 5.1 Donald J Hughes, in 1955 (left) and Jack Chernick, in 1958 (right), to whom the Brookhaven high-flux reactor can be primarily credited.

During the years the HFBR was taking form, the BGRR experimental programme went on much as before. Nearly a decade had passed since Hughes's original meeting with Hayworth and only a few of those then active in neutron research—Lester M Corliss, B Chalmers Frazer, Julius M Hastings, Harry Palevsky and Vance L Sailor—were still at the BGRR. Only two, Corliss and Hastings, would continue in the neutron field well into the HFBR era. A good part of the responsibility for planning the HFBR experimental programme therefore fell on the shoulders of more recent arrivals. Robert Nathans was one. His first contact with neutron research had been as a visitor to the Laboratory in 1955 when he helped Clifford G Shull build a polarised neutron spectrometer at the BGRR. (Shull had moved from Oak Ridge to Brookhaven while awaiting completion of the new Massachusetts Institute of Technology reactor.) Another new arrival was Gen Shirane who began working at the BGRR in 1956 to investigate ferroelectric structures. Finding polarised neutron spectroscopy more to his liking than crystallography, he decided to join Shull and Nathans in a study of magnetic form factors. Later, the measurements completed, all three moved on; Shull settling down, as planned, at MIT, while Nathans and Shirane, after spending time at other institutions, returned eventually to Brookhaven, Nathans in 1960 and Shirane in 1962. Walter C Hamilton arrived at the Laboratory in 1956: a scientific prodigy only 25 years of age, he came to establish a programme of crystallographic studies at the BGRR. In 1957, Robert E Chrien joined the staff. To his surprise, he was immediately packed-off to Chalk River by Hughes to work on the NRU fast chopper. I arrived in 1963 to join a group organised by

Vance Sailor to study the interactions of polarised neutrons with polarised nuclei.

By then, preparations for the forthcoming HFBR experimental programme were already in progress. Much of the work involved designing and building upgraded versions of equipment then in use at the BGRR. The only really new departures in instrumentation—aside from some state-of-the-art applications of computers to data handling and instrument control problems—were in the area of inelastic scattering: three nearly identical triple-axis spectrometers were designed and built between 1962 and 1965 and work was begun on a two-rotor, slow chopper time-of-flight system intended for use at the HFBR with a liquid-hydrogen cold neutron moderator.

Chrien, who had assumed responsibility for the fast chopper programme when Hughes died, was the first to begin operations at the HFBR. His newly built time-of-flight system was installed there in mid-1966. By the end of the year most of the other BGRR experimental groups had followed. Soon afterwards the BGRR was shut down. In 15 years its research staff had compiled a remarkable record: Hughes's early studies of neutron optics, Corliss and Hastings's exploratory investigations of magnetic structures, Palevsky and Hughes's discovery of critical magnetic scattering, Ramsey's definitive measurement of the magnetic moment of the neutron, Shull, Nathans and Shirane's pioneering work on magnetic form factors, the first direct observation of rotons in superfluid helium made by Palevsky and his collaborators . . . the list goes on and on. But they had little time to think of past achievements as 1966 drew to a close; it was the immediate future that occupied their minds.

After the briefest of shakedown periods, the HFBR experimental programme quickly hit its stride. Chrien's Fast Chopper Group, taking advantage of the recently introduced lithium-drifted germanium gamma ray detectors, now found itself in a position to investigate capture gamma ray cascades in unprecedented detail and soon settled down to an extended programme of nuclear level studies that was to be their central interest for many years to come (Chrien 1980). Hamilton and his collaborators wasted no time in addressing a long-accumulated backlog of crystallographic problems that had been outside their range at the BGRR. A definitive study of decaborane—to mention only one of many examples—dates from the early HFBR years. Before his untimely death at age 41 in 1973 Hamilton had also begun to develop an interest in metal–hydrogen bonds (La Placa et al 1969). The programme he established in this field has since been carried on, along with work in other areas, by his successor, Thomas F Koetzle (Hart et al 1979). In 1970, Benno P Schoenborn formed a second crystallographic group at the HFBR. Schoenborn, a biophysicist, was one of the first to see the advantages of using neutrons as structural probes of biological macromolecules. Over the years he and his collaborators have

Figure 5.2 The experimental floor at the Brookhaven high-flux reactor, photographed (above) in 1964 and (below) in 1982.

investigated a wide variety of molecular structures. Their study of the arrangement of macromolecules in the bacteria *E. coli* is one of many made at the HFBR in this active and important field (Engelman *et al* 1975).

Corliss and Hastings first started exploring magnetic materials at the BGRR soon after it became operational in 1950 and continued to work on magnetism after moving to the HFBR. Now, however, using one of the new triple-axis spectrometers, they could probe dynamical as well as structural properties. Their classic study of the dynamical response of rubidium manganese fluoride, a prototype Heisenberg antiferromagnet, was carried out during the early HFBR period (Tucciarone *et al* 1971). Later they went on to examine many other interesting aspects of magnetic

Figure 5.3 An external view of the Brookhaven high-flux reactor in 1964, the year before completion.

behaviour: studies of the effects of inducing a staggered magnetic field in the antiferromagnet dysprosium aluminium garnet and of the details of the spin-flop process in manganese dichloride tetrahydrate can only suggest the breadth and scope of their wide-ranging programme.

Nathans and Shirane assumed joint responsibility for the other two triple-axis spectrometers. At first, magnetic systems such as the ferromagnets iron and nickel, the antiferromagnet manganese fluoride and the two-dimensional Heisenberg antiferromagnet potassium nickel fluoride were their primary concern (see, for example, Collins *et al* 1969). Although a broader spectrum of interests gradually developed after Nathans' departure in 1968, work on magnetism continued to be an important part of their research programme. In the years that followed, the properties of the one-dimensional Heisenberg antiferromagnet dichloro-bis-pyridine copper II were explored along with numerous prototype systems such as the isotropic Heisenberg ferromagnets europium oxide and europium sulphide and various diluted two- and three-dimensional antiferromagnets. Mention should also be made of recent work done on a group of remarkable ternary compounds in which magnetism is found to coexist with superconductivity (Thomlinson *et al* 1982).

The two-rotor slow chopper came on-line in 1967, after a long and arduous period of development. Proving difficult to maintain, it was replaced after a few years by a triple-axis spectrometer. During its brief lifetime as a condensed matter probe, however, it was employed to make an extended study of the pressure and temperature dependence of roton excitations in liquid helium (Dietrich *et al* 1972) and for a pioneering survey of crystalline electric field transitions in rare-earth metallic compounds (Turberfield *et al* 1971).

John D Axe, who came to Brookhaven in 1969, was instrumental in directing the HFBR research programme into many new and unexplored areas. Of particular interest are his studies of the role of phonons in certain special classes of phase transitions: the soft-mode-initiated structural transformations and the superconducting transitions. The work he and Shirane have done on these systems (Axe and Shirane 1973a,b) provides a remarkably detailed picture of an important group of phonon-mediated processes. Another often-cited experiment in which Axe played an influential part involved the investigation of the subtle lattice distortions brought about by charge-density-wave instabilities (Moncton *et al* 1975).

Stephen M Shapiro became a member of the HFBR research staff in 1971. After first focusing his attention on the difficult question of how the cubic-to-tetragonal phase transitions of the perovskites strontium titanate and potassium manganese fluoride influence their microscopic dynamical properties he branched out into other areas and, in time, broadened the scope of magnetic studies at the HFBR even further by developing an interest in the so-called 'spin glasses', a curious group of materials in which the magnetic ground state is one of frozen-in disorder. Over the years he and his collaborators have made numerous studies of the collective magnetic excitations of a special subset of these systems—those undergoing ferromagnetic-to-spin-glass transitions—and have done much to characterise this unusual type of phase transformation (Aeppli *et al* 1984).

My own work at the HFBR through 1971 with ferromagnets and liquid helium has already been mentioned in relation to other members of the staff. Since then I have found myself increasingly drawn to studies of the behaviour of weakly bound monolayer gas films adsorbed on graphite basal plane surfaces (Taub *et al* 1977). Often behaving like independent two-dimensional entities, these films have opened a wide range of surface structures, dynamical responses and phase transitions to experimental investigation.

No account of the HFBR would be complete without paying tribute to the important part visiting scientists have played in the research programme. Such stalwarts of the scientific community as Jens Als-Nielsen, Robert J Birgeneau, Roger A Cowley, David E Moncton, Tormod Riste, Clifford G Shull—to name only a few of the many who have used the facilities of the HFBR—have been the driving force behind a number of experiments which stand out as highlights of the past two decades.

1985 marks the twentieth anniversary of the HFBR. Bolstered, in 1980, by the addition of a liquid hydrogen moderator and, in 1982, by a power increase from 40 to 60 MW, it is still one of the world's outstanding steady-state research reactors—a testament to the extraordinary competence and foresight of its creators and to the dedication and skill of its operations and research staffs.

References

Aeppli G, Shapiro S M, Maletta H, Birgeneau R J and Chen H S 1984 *J. Appl. Phys.* **55** 1628
Auerbach T, Chernick J, Juliens J, Lellouche G, Zinn W and Associates 1958 *Proc. Second Int. Conf. Peaceful Uses of Atomic Energy* vol 10 p 60
Axe J D and Shirane G 1973a *Phys. Today* **26** 32
—— 1973b *Phys. Rev. Lett.* **30** 214
Chrien R E 1980 *Trans. NY Acad. Sci.* II **40** 40
Collins M F, Minkiewicz V J, Nathans R, Passell L and Shirane G 1969 *Phys. Rev.* **179** 417
Dietrich O W, Graf E H, Huang C H and Passell L 1972 *Phys. Rev.* A **5** 1377
Engelman D M, Moore P B and Schoenborn B P 1975 *Proc. Nat. Acad. Sci. USA* **72** 3888
Hart D W, Teller R G, Wei C-Y, Bau R, Longoni G, Campenella S, Chini P and Koetzle T F 1979 *Angew. Chem. Int. Ed. Engl.* **18** 80
La Placa S J, Hamilton W C, Ibers J A and Davison A 1969 *Inorg. Chem.* **8** 1928
Moncton D E, Axe J D and di Salvo F J 1975 *Phys. Rev. Lett.* **34** 734
Taub H, Carneiro K, Kjems J K, Passell L and McTague J P 1977 *Phys. Rev.* B **16** 4084
Thomlinson W, Shirane G, Lynn J W and Moncton D E 1982 *Superconductivity in Ternary Compounds II* ed M B Maple and O Fischer (Heidelberg: Springer)
Tippe A and Hamilton W C 1969 *Inorg. Chem.* **8** 464
Tucciarone A, Lau H Y, Corliss L M, Delapalme A and Hastings J M 1971 *Phys. Rev.* B **4** 3206
Turberfield K C, Passell L, Birgeneau R J and Bucher E 1971 *J. Appl. Phys.* **42** 1746

5.2 High Flux at Oak Ridge

J W Cable

Oak Ridge National Laboratory, USA

At the Oak Ridge National Laboratory (ORNL), the High Flux Isotope Reactor (HFIR) went critical in 1965 and reached full design power in the autumn of 1966. The HFIR is a beryllium-reflected, light-water cooled and moderated reactor with a flux-trap design. It was designed primarily for the production of transplutonium isotopes for use in the heavy-element chemistry programme, but several other experimental facilities were provided. These include four horizontal beam tubes which originate in the beryllium reflector and four slant tubes located adjacent to the reflector. The thermal neutron flux at the entrance to the horizontal beam tubes $(1.3 \times 10^{15}$ neutrons $cm^{-2}s^{-1})$ is an order of magnitude higher than was previously available at ORNL. This flux increase, in conjunction with other technological developments, naturally afforded new research opportunities, some of which led to the expansion of existing programmes while others resulted in the introduction of new research programmes. In this section, we give a brief description of this changing scene in an attempt to provide an impression of the HFIR's impact on neutron scattering research at ORNL.

In the pre-HFIR period there were only two neutron scattering research groups at ORNL; the physics group, led by E O Wollan, which had a strong programme in magnetism and the chemistry group, directed by H A Levy, which had well established programmes in crystallography and the structure of liquids. These programmes continued, with some modifications, at the HFIR. The magnetism programme of the physics group was, and still remains, directed toward understanding the fundamental interactions in solids. Before the HFIR came on-line, this group was primarily concerned with the determination of magnetic structures but they also performed studies of paramagnetic scattering, critical scattering, magnetic form-factors and magnetic moment distributions in ferromagnetic alloys. Measurements were made with unpolarised neutrons on two-axis diffractometers located at the Oak Ridge Research Reactor (ORR). By the time of the instrument design stage for the HFIR, the use of polarised neutron beams for magnetic studies was becoming increasingly important. Polarised neutron diffractometers had been constructed at the Brookhaven National Laboratory and at the Massachusetts Institute of Technology where they were in use for the precise determination of magnetic form factors. Polarised neutrons were being used at Harwell in studies of the magnetic diffuse scattering from disordered ferromagnetic alloys and the group at Kjeller had demonstrated the use of polarised neutrons in separating spin-wave scattering from other inelastic scattering effects. In view of these developments, it was clear that polarised neutrons should be available at the HFIR and the first two instruments were designed with that capability.

The first polarised beam study at the HFIR was an extensive series of experiments on the polarisation effects in various types of scattering. This led to the development of the polarisation analysis technique (Moon *et al* 1969) in which a triple-axis spectrometer with polarisation-sensitive crystals for both the monochromator and the analyser is used to determine the energy transfer, the momentum transfer and the change in neutron spin resulting from a scattering process. A variety of spin-dependent scattering effects was demonstrated which showed the value of this new technique in

studies of paramagnetic scattering, spin-wave scattering, nuclear incoherent scattering and Bragg scattering in complex antiferromagnets. Since
these initial experiments, polarised neutron beams have been found so
useful in magnetic studies that the HB-1 triple-axis spectrometer has been
used almost exclusively in the polarised mode. Studies that have been
carried out include diffuse scattering from ferromagnetic alloys, form-
factors of the induced magnetisation density in paramagnets, magnetovib-
rational scattering, forbidden magnons near T_c and spin-pair correlations in
spin glass alloys.

The chemistry group makes use of both X-ray and neutron diffraction
methods in their studies of crystal and liquid structures. In the time period
leading up to HFIR, the field of crystallography was moving toward ever
larger unit cells containing larger numbers of atoms; i.e. crystal structure
determinations required the collection of intensity data for an increasingly
larger number of Bragg reflections. This group was in the forefront of the
efforts toward the automation of diffractometers for rapid data collection
and in the development of computer methods for data handling and for the
refinement of structural parameters. They designed one of the first
computer-controlled, four-circle, X-ray diffractometers and this was
placed in operation two years before the HFIR became available. A similar
neutron diffractometer was then built and installed at the HFIR. This
provided the capability of rapid, automatic data collection for crystal
structure studies with better Q resolution and shorter counting times than
was previously possible.

In this same time period, the liquid structure programme was involved in
studies of the structure of molecular liquids including the most important
and interesting of liquids, water. Unfortunately, water is also one of the
more difficult liquids for structural studies. The oxygen–oxygen pair
correlation can be determined by X-ray techniques, but the oxygen–
hydrogen and hydrogen–hydrogen pair correlations must be taken from
neutron data which require not only isotopic substitution but also large
corrections for dynamic effects and for incoherent scattering from hydrogen. With the advent of the position sensitive detector (PSD) in the early
1970s and with the higher flux of the HFIR, it became possible to collect
sufficiently accurate data so that these corrections could be made with a
satisfactory degree of confidence. A liquids diffractometer equipped with a
one-dimensional PSD was installed in 1976 by A H Narten. This unit was
initially used by Thiessen and Narten (1982) to study the structure of liquid
water and is now used in studies of solvation effects in aqueous solutions.

We now turn to the new programmes associated with the HFIR. The first
of these was a neutron spectrometry group organised by M K Wilkinson in
1963–64. By this time, the field of inelastic neutron scattering had already
reached an advanced stage. It had received a tremendous boost from the
development of the triple-axis spectrometer and the constant-Q and

constant-E methods of data collection and was growing rapidly both in the number of practitioners and in the range of applications. With the completion of the HFIR only two years away, the neutron spectrometry group quickly assembled a conventional triple-axis spectrometer at the ORR in 1964 (see figure 5.4). This was later moved to the HFIR. Meanwhile, the design and construction of the new triple-axis spectrometers was underway. These were designed for high flexibility with a monochromator arrangement that allowed rapid changeover between polarised or unpolarised neutron beams (Wilkinson *et al* 1968) These spectrometers were installed in the spring of 1967 (see figures 5.5 and 5.6). Instrumentation for the neutron spectrometry group was completed by the installation of a magnetically pulsed time-of-flight spectrometer (Mook and Wilkinson 1970) and, finally, another triple-axis spectrometer in 1978.

The group effort was directed toward studies of the dynamic properties of solids. Neutron techniques are particularly well adapted to such studies because the energies and momenta of thermal neutrons are comparable to those of the characteristic excitations in solids. The frequency spectra of lattice vibration or spin-wave modes contain essential information on the interatomic forces and electronic structures of the material, much of which can be obtained by no other experimental method. In the area of lattice vibrations in metals, much of the interest is related to conduction electron

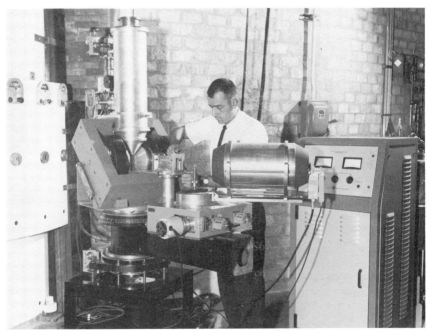

Figure 5.4 The first Oak Ridge triple-axis spectrometer attended by R M Nicklow of the neutron spectrometry group.

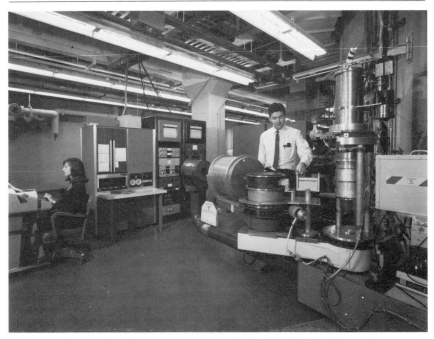

Figure 5.5 The triple-axis spectrometer at the HFIR. H G Smith of the neutron spectrometry group is at the spectrometer and S P King, who did most of the programming for spectrometer control, is seated at the console.

screening of the interatomic interactions, i.e. the phonon frequency spectrum of a metal reflects both the ion–ion and the ion–electron interactions. Because of this sensitivity to electronic properties, phonon anomalies are observed that can be related to superconductivity, incipient lattice instabilities and Fermi surface density-of-states effects. One of the early studies at the HFIR showed a pronounced softening of certain acoustic phonon modes in the transition metal carbides that could be correlated with the occurrence of high-T_c superconductivity (Smith and Glaser 1970). Another early result was the first observation of a localised vibrational mode in a mass defect system (Nicklow *et al* 1968). Later studies included layered pseudo two-dimensional systems, lattice instabilities and charge density waves. Much of this work is surveyed in recent review papers, such as Smith and Wakabayashi (1977) and Nicklow (1979).

In the area of spin-dynamics there is a rich diversity in the type of magnetic system subject to experiment. Part of this diversity arises from the degree of localisation of the electrons responsible for the magnetism and part is due to the seemingly endless variety of long-range and short-range order produced in magnetic systems by the interactions between the spins. Magnetic inelastic neutron scattering determines the magnetic response of the system in terms of the same generalised susceptibility

Figure 5.6 The polarised-beam, triple-axis spectrometer at the HFIR. S P King, R M Moon and J L Sellers inspect the RF power supply for the spin flipper.

function in which the theories are formulated. The experiments therefore provide a most stringent test of the theories. The early spin-dynamics studies at the HFIR were concerned with the spin-wave spectra of the itinerant electron ferromagnets, Ni and Fe (Mook *et al* 1969)†. These experiments revealed new effects, one of which was a strong damping of the propagating spin-wave modes due to interactions with single-particle excitations. Another observation was the apparent persistence of spin-wave-like modes at temperatures above T_c. These experiments stimulated considerable theoretical effort towards understanding the generalised susceptibilities of itinerant electron systems.

Other new programmes developed at the HFIR make use of small angle neutron scattering (SANS) techniques. SANS is generally regarded as a diffuse scattering technique covering a small-Q range that corresponds to fluctuation in scattering density on the scale of 10–1000 Å. However, the technique can also be used for Bragg scattering from systems with large periodicities. A programme using the latter method was initiated at ORNL in 1975 to study the properties of flux-line lattices in type-II superconductors (Tasset *et al* 1976). These materials retain their superconductivity in high magnetic fields due to lines of magnetic flux density that penetrate the

† See also Mook and Nicklow (1973), Mook *et al* (1973) and Lynn (1975).

superconducting matrix. Within certain limits of the applied magnetic field, these flux lines form a two-dimensional lattice which can be examined by small-angle neutron diffraction to determine such properties as the structure, the lattice parameter and the form-factor of the magnetic induction. The ORNL programme, under the direction of D K Christen (1980), has been concerned with properties of pure Nb and the A-15 type-II superconductors. The neutron measurements are made on a double, perfect-Si crystal diffractometer.

A more recent addition to the instrumentation at the HFIR is a national user-oriented SANS facility that was constructed in 1980. This is part of a joint Oak Ridge National Laboratory–National Science Foundation–Department of Energy project that operates in a full-time user mode under the direction of W C Koehler. The machine uses a 65×65 cm^2 two-dimensional PSD which can be translated along the neutron flight path to cover the Q range from 0.003 to 0.6 Å^{-1}. Most of the experiments employ the diffuse SANS technique which has an extremely wide range of application to problems in materials science, polymer science, physical chemistry and biology. The polymer community has been the major user of the facility so far, but there is increasing usage from the other disciplines.

The most recently developed programme making use of the HFIR is the USA–Japan Cooperative Program on Neutron Scattering. The programme is centred around a wide-angle neutron diffractometer that was installed at the HFIR in 1984. This diffractometer uses a curved one-dimensional PSD to cover a wide angular range and can be used either for polycrystalline samples or, in inclination geometry, for single-crystal diffraction. The data acquisition system is capable of handling time-resolved diffraction data in either a step-mode or a time-slicing mode. This collaboration has also led to the acquisition of sophisticated auxiliary equipment including a dilution refrigerator, furnaces for rapid temperature changes and high pressure cells, all of which allow studies to be performed under extreme environmental conditions. So far the programme has been directed only towards problems in solid state physics but will eventually expand into the fields of metallurgy, ceramics and chemistry.

We have described some of the changes in neutron scattering research at ORNL associated with the advent of the high-flux reactor. These were significant changes ranging from the expansion of existing programmes in physics and chemistry to the development of new research programmes in inelastic and small-angle scattering. We recognise that these are only the outward or more visible changes and that there were many other less obvious advances in neutron scattering research associated simply with experimental feasibility. The experimentalist in neutron scattering is usually pushing the technique to one limit or another. Any increase in source flux allows an improvement in intensity or resolution or perhaps in the accessible range of Q and ω space. This broadening in the scope of experimental feasibility is the real impact of the high-flux reactor. At Oak Ridge, the HFIR is a third-generation neutron source; we are looking forward to the next generation.

References

Christen D K, Kerchner H R, Sekula S T and Thorel P 1980 *Phys. Rev.* B **21** 102
Lynn J W 1975 *Phys. Rev.* B **11** 2624
Mook H A, Lynn J W and Nicklow R M 1973 *Phys. Rev. Lett.* **30** 556
Mook H A and Nicklow R M 1973 *Phys. Rev.* B **7** 336
Mook H A and Wilkinson M K 1970 *Instrumentation for Neutron Inelastic Scattering Research* (Vienna: IAEA) p 173
Moon R M, Riste T and Koehler W C 1969 *Phys. Rev.* **181** 920
Nicklow R M 1979 in *Neutron Scattering* ed G Kostorz (New York: Academic) p 191
Nicklow R M, Vijayaraghavan P R, Smith H G and Wilkinson M K 1968 *Phys. Rev. Lett.* **20** 1245
Smith H G and Glaser W 1970 *Phys. Rev. Lett.* **25** 1611
Smith H G and Wakabayashi N 1977 in *Dynamics of Solids and Liquids by Neutron Scattering* ed S W Lovesey and T Springer (Heidelberg: Springer) p 27
Tasset F, Christen D K, Spooner S and Mook H A 1976 *Conference on Neutron Scattering* CONF-760601-P1 Oak Ridge National Laboratory, USA p 481
Thiessen W E and Narten A H 1982 *J. Chem. Phys.* **77** 2656
Wilkinson M K, Smith H G, Koehler W C, Nicklow R M and Moon R M 1968 *Neutron Inelastic Scattering* vol. II (Vienna: IAEA) p 253

5.3 International Science

5.3.1 The birth of the Institut Max von Laue–Paul Langevin in Grenoble

H Maier-Leibnitz, Munchen, West Germany

It is well known that the High-Flux Reactor in Grenoble has its roots in an effort within the Organisation for Economic Cooperation and Development (OECD) to find a project big and important enough to be carried out as a European or European–American enterprise. There was a committee composed of scientists from the OECD countries, with L Kowarski as chairman. The High-Flux Reactor turned out to be the most hopeful candidate, and rather detailed plans were made both for the reactor and an experimental programme. Among the Europeans, scientists from Harwell and from Saclay were, of course, leading, because of their successful research reactors.

The contributions of P Egelstaff must be mentioned especially. However, when it came to decisions, the British Government would not take part, and this finally led to the plan to build and operate the reactor between France and Germany. The reactor was redesigned, mainly between Saclay, Karlsruhe and Grenoble, with Robert Dautray as the first *Chef de Projet*. For the planning of experiments, there was, in addition to what had been done before, more influence now from the newer research reactor stations in Germany, with advice from our friends in other countries in Europe, and the United States and Canada. The decisions, however, had to be made by the French and Germans alone, recommended by a Scientific Committee and agreed to by the Comité directeur.

It appears that the initial limitation of the project to two countries had its beneficial side. The proposals for experimental equipment had so far represented the standard that had been achieved at the first leading reactors in the US, in Canada, in Great Britain, and in France. Now a second generation of experiments came up, starting mainly but not only in Munich. It is not always easy to convince even experienced colleagues of an idea or a new method that has not been used before. Starting a project in a small group where all the participants, including those who control the budget, know each other well and find time for complete discussions, may lead to an atmosphere of mutual trust that will favour innovation. We were very lucky in this respect. Internationally, the reactor users were a small group bound together by common interests. From the beginning, i.e. after the first Geneva Conference on the Peaceful Uses of Atomic Energy, we had received information and help at all the reactor stations we visited, in the United States, in England, in France. And when we started doing things which had not been done elsewhere (starting, it is true, with Mössbauer's discovery which had nothing to do with a reactor), there was a flood of visitors and of invitations. In France, we were helped by the memory of friendly and important relations which originated during and after the war between Frédéric Joliot and Wolfgang Gentner. We were greatly helped by the two organisations that took care of the project, the Commissariat à l'Energie Atomique in France and Gesellschaft für Kernforschung Karlsruhe on the German side. They again had the support of the governments which favoured a purely scientific project at a time when the political situation was somewhat strained in the European Community. And in Grenoble, Louis Néel, who played a leading role in all of science there, brought to us the help of the Centre d'Etudes Nucléaires de Grenoble, of the institutes of the Centre National de la Recherche Scientifique, and the University. And behind all this was the spirit of CERN, the fabulously successful European cooperation in high-energy physics.

These were the favourable conditions under which the project was started. What have been our own contributions to ensure its future success? I shall try to list them, even if I am not sure that my list is complete, or correct in all points. The first is that after the government treaty had been signed, operations were started in Grenoble by the two directors (Bernard Jacrot on the French side) and one secretary, leaving no doubt about who was responsible. The second point was that the future director of administration arrived only a few weeks later. This was important because for whatever we decided, we had a choice of doing it the French or the German way, or to invent something of our own. Another point was that the reactor project was separate, not under the responsibility of the directors, but with a close cooperation between us and the *Chef de Projet* who did everything to respect the wishes of the experimenters.

Probably the most important decision concerning the future organisation was that the reactor was to be used mainly by scientists from outside the Institut Laue–Langevin, mainly from the two countries but not excluding others. We have had a steady participation of the latter of the order of 18 per cent, with no cost to them and with no objections from our Comité directeur. They have not only profited from this but have given us invaluable advice and help. But our internal decisions were even more comprehensive and stringent. Guests from the two countries were not only to work in Grenoble free, but their local costs for additional equipment and even travelling costs could be paid by the Institute under certain circumstances, because we knew that such things could create great difficulties at their home universities. Contracts for scientists in Grenoble were to be limited to five years, with a few exceptions, mainly for those whose duties included more than the usual amount of service for others. This has been modified later to allow leading scientists, for instance from theoretical physics, to come to Grenoble on a permanent basis.

For the scientists at Grenoble, the need to work mainly for a future when others would profit by their efforts, certainly demanded some idealism and unselfishness. On the other hand, developing optimum equipment for all future experiments was a challenging task, rewarding in itself, and made even more attractive because those who developed new methods would also be the first to use them. What helped again was that nearly all experiments in the domain of bound matter (in contrast to nuclear physics about which we shall not speak here) follow the same fundamental pattern: a neutron beam defined in phase space (real space plus three coordinates of momentum) impinges upon a target, and what is measured is the probability of detecting a neutron scattered into some other element of momentum space. This simple and universal process allows an effort to develop relatively few instruments for all possible applications. For magnetic measurements, the spin direction of the neutron before and after scattering must be determined, too.

The state of the art as it had been developed the world over and mainly in the USA, in England and in France, was at our disposal through our own visits there, through the work of the OECD committee including Egel-staff's contributions, and now through the help from Saclay and CENG, where some of our instruments were developed. From the reactor project group came a design which furnished slightly higher neutron intensity than our main existing competitor, the High-flux Reactor at Brookhaven. But there were two major improvements. One was a so-called 'cold source', consisting of liquid heavy-hydrogen near the reactor core. It yielded an intensity gain of more than 100 for all experiments that could be done with very slow neutrons. At the other end of the thermal neutron spectrum, the intensity of faster neutrons was increased by a hot source of graphite. The safe design of these sources was a major engineering feat.

Other innovations had their origin in Munich, or at Jülich where the early Munich work had been continued and perfected. The research reactor in Munich, the first in Germany, was small, but it came at a time when an extraordinary number of students in physics wanted training by research. Five years earlier, we had started to establish research in nuclear physics, and now there were large teams ready to do all kinds of experiments at the reactor. We tried to do things which had not been done elsewhere before, and to develop new or improved methods, using simple ideas. Some of the young scientists from Munich had since gone to Jülich when the Kernforschungsanlage there was started, and so together we were prepared to take an active part in the development of instruments for the Grenoble reactor.

The first contribution was establishing neutron guide tubes. There was some opposition at the time, but they have proved to be an invaluable tool since. Any neutron that hits, say, a nickel surface while it has a velocity component of not more than $6\,\mathrm{m\,s^{-1}}$ normal to the nickel surface, is totally reflected because it cannot penetrate into the nickel. Therefore, such neutrons may be guided between two nickel mirrors over long distances. And if the mirrors are slightly bent, the thermal neutrons can be separated from the other reactor radiations which cannot be thus deflected, and instruments may be operated far from the reactor with hardly any background. The capacity of the reactor for simultaneous experiments has been increased by a factor of about 2.5 by establishing ten neutron guide tubes in a hall outside the containment shell of the reactor.

The reactor in Munich had low power, so we always had intensity problems compared with the larger reactors where the important work was done. This led us into careful studies of the relation between intensity and resolution, using phase space representations and Liouville's theorem. This later became important in optimising instruments for the Grenoble reactor. But our first application was a very simple one: when neutrons are diffracted normally on a lattice plane, their velocity is nearly constant within a not very small solid angle around this direction. This was used in Munich and later at Jülich to do certain measurements with a resolution that was more than a hundred times better than usual. This of course became a much-used intrument in Grenoble. In Munich, a time-of-flight version was built for back scattering from powders.

While the reactor was under construction, we anticipated a great demand for neutron diffraction on small, mostly organic crystals mainly because neutrons can see protons directly, which X-rays cannot. We therefore invented two instruments which made limited use of the fact that neutron beams have a wide velocity spectrum: a bent monochromator crystal focuses neutrons from a modified Laue pattern of diffracted neutrons. They may be detected either photographically or with a multi-detector system of 100 movable helium counters. The first method worked successfully but was not much used later on. The second one, the so-called 'hedgehog' spectrometer, was later abandoned mainly because we had underestimated the problem of the background around the peaks. A new version of this may come up in the near future. But for the time single-crystal diffraction had to be done on conventional instruments. At the Brookhaven High-flux Reactor, our older competitor, the demand for single-crystal work seems to have been more urgent. They have since developed an instrument which is superior to ours.

There were quite a number of conventional instruments at Grenoble, for diffraction on liquids, for polarised neutrons, and especially for inelastic scattering. They were designed and constructed with the help of our friends in Grenoble, at Saclay and at Jülich, with help even from our colleagues from Risø and from Oak Ridge. They included some efforts at 'focusing' (increasing phase space volumes, see above). We had anticipated a future need for high-resolution work in inelastic scattering, for instance on the widths of phonon lines which give indications on non-linearity. Therefore, we had proposed a rather futuristic instrument where the monochromators and samples moved on air cushions; everything was variable and computer-controlled, and complex, curved or strained monochromators were fore-seen. This instrument has not so far been used as planned. Few people are willing to carry out alone ideas which are not their own and to overcome all the difficulties which only show up during the actual work. However, the instrument is used with multidetectors which have since gained in feasi-bility and in importance. And the development of monochromator crystals of various materials and properties has become a great effort and a great success at the ILL, partly in cooperation with Jülich.

The next instrument originated from a research problem which we had in Munich: lattice defects that were produced by low temperature irradiation of metals by neutrons, could best be investigated by observing diffuse scattering of slow neutrons below the Bragg limit. This kind of work was continued at Jülich and led to a number of well used instruments in Grenoble, employing multiple detectors.

The most successful instruments were those for small-angle scattering.

Small-angle scattering of neutrons was first tried in Great Britain because it promised evident advantages over conventional X-ray small-angle scattering. We tried it in Munich, adding a most trivial but very helpful feature: the intensity of a neutron beam is proportional to the area of the target which it covers. When all dimensions of an instrument including the target are increased by a factor of, say, 20, the resolution remains the same but the intensity goes up by a factor of 400. The main instrument in Grenoble is more than 100 m long. In addition, it has beautiful multidetectors which were developed by LETI, a section of CENG. Their help has been a great contribution to the success of ILL. In all cases where large samples are not an obstacle, neutron small-angle scattering can now be done with the same resolution as is obtained with X-rays. Potentially all users of X-ray scattering immediately saw the new possibilities that were offered by the use of neutrons. Therefore, small-angle scattering is now the most sought-after method in Grenoble, with long waiting times and even selection between proposals or short beam times, which is never a good thing when all the applicants are good groups who know more about their project than the referees can know. We were quite surprised by this success because we had no idea how important small-angle scattering was in several fields, especially in biology.

My five years in Grenoble were over when the reactor was finished and the instrument programme described here was nearly complete. The British participation had by this time been decided but came into effect a little later. Looking back, a director will always ask himself questions. Did we employ our time well? Was the great effort (the ILL employs 400 persons) which our governments allowed us to make, justified? Was it good policy not to aim at scientific autonomy for ILL in all the fields in which the reactor was to be used? Each of us had his own small field and there were experts in Grenoble, like E Bertaut, who had been of the greatest help, but we had to learn a lot from each other and from our guests, and we had to rely on them almost completely for the scientific future of the institute. Did we encourage the future users? Did we offer them enough help to enter a field that was unfamiliar to most of them? And in this connection is it reasonable to believe that one can develop for others the instruments and methods that they will eventually find really useful for their work?

Satisfactory answers to all of these questions cannot be given, for it is just the same with the organisation of research as it is with research itself; it could always be better. However, now more than a dozen years later, we can see a little more clearly. The ILL has been a success, and it will continue to be a success for quite some time. The reasons for this are, I think, that the effort was adequate for the task, that we made it a customer's reactor, and that we gave our guests freedom and help. The instruments we planned and built may not all have fulfilled all the tasks they were meant for, but they gave a reasonable start and above all, they set the level for all the future developments including the *'deuxième souffle'*. There has been fundamental progress like Mezei's sensational invention of the neutron spin echo method, or the development of polarising mirrors, but most of the many improvements are improvements using established principles.

As I wrote this article, I had a deep feeling of gratitude towards all the many people who were connected with the ILL project. One person cannot do much, many can achieve more, but real success comes through a spirit of cooperation, and I suppose everybody will agree that this was a special feature of our enterprise.

My special thanks are due to all those who were meant when I used the word 'we' in this text. I may have overemphasised their (our) contributions compared to those of others because I knew them better, but I want to say that 'we' does not mean 'I' but they who did the work and who overcame all the difficulties that arise when one tries to make an idea work.

Figure 5.7 The location of the Institut Laue–Langevin at Grenoble near the confluence of the rivers Isère and Drax and about 1½ miles from the railway station.

Figure 5.8 Professor H Maier-Leibnitz, first Director of the Institut, with his co-Director, B Jacrot, during the second meeting of the ILL Steering Committee on 28 October 1968.

Figure 5.9 The construction site of the Grenoble HFBR in summer 1970.

Figure 5.10 Professor R Mössbauer, Director of the Institut, speaking on the occasion of the reception of the UK as a new member of the ILL in January 1973.

Figure 5.11 The guide tubes leading from the thermal column of the reactor to the remote guide-hall at the ILL (thermal guides underneath, 5 cold guides above).

Figure 5.12 A complex and advanced ILL instrument: D7, for elastic and quasi-elastic diffuse scattering.

5.3.2 Twenty years (or so) of Science at the ILL: a view from the Secretariat

G A Briggs, Institut Laue–Langevin, Grenoble, France

On January 1987 we celebrated the 20th anniversary of the signing of an agreement to build and operate a high-flux reactor entirely devoted to the production of neutron beams. Noble ideals were embodied in the document, the single aim being the advancement of science in the spirit of international collaboration.

The principles were clear, their ultimate fulfilment less evident at the time, and by 1972, when the reactor went critical for the first time, it was decided to create a Scientific Secretariat for the management of the increasing interest which was being shown by neutron scientists wishing to use the most powerful beams which had yet been made available to the scientific commmunity. A Scientific Council had already been appointed from the French and German scientific community (the UK became an equal partner in January 1973) and it became necessary to appoint adjudicating sub-committees of the Council in order to assess the excellence of experimental propositions to use the instruments that were already being installed. There was one instrument available in 1972, IN2, a triple-axis spectrometer, and after 5 years of intensive building and planning we had the first experiment under way (spin wave measurements in the one-dimensional ferromagnet $CsNiF_3$) eventually, the first scientific paper (Steiner and Dorner 1973 *Solid State Commun.* **9** 537). The ILL was in business.

At the time of this first experiment, the Institute was in direct competition with numerous domestic European and other reactor stations, as the late 1960s and early 1970s had heralded a virtual explosion of interest in the powerful experimental tool provided by monochromatic neutron beams. In many cases, neutron beam research facilities were parasitic on reactors designed mainly for other purposes such as materials testing, isotope production, loop verification etc, and with the changing pattern of financial support both for reactors and future reactor design, 1980 saw the beginning of a gradual decline in the availability of instrumentation for pure and fundamental neutron research. It could be argued that this was also because conventional beam reactors had, broadly speaking, reached their limit in tems of neutron flux production and would be eventually replaced by accelerator based pulsed neutron spallation sources for such scientific applications as were suitable and for those experiments able to benefit from the projected flux increases.

The Institut Laue–Langevin had at this time, however, a custom-built high-flux steady-state beam reactor operated as a central facility and it largely survived the rationalisation that was going on elsewhere. Additionally, the policy of successive Directors had been to have, and to plan for, an extremely comprehensive instrument portfolio using long wavelength cold, very cold and ultra cold neutrons, thus transforming a major part of the Institute to a unique facility and wholly complementary to future pulsed spallation sources.

With over a decade of experience under fully operational conditions and offering the use of some 60 instruments both fully scheduled and special beam, the Institute quickly acquired a reputation as a centre of excellence unequalled in the world. It is now recognised as a pattern both for international scientific cooperation and, above all, for the economic use of increasingly scarce research funds. Among the member countries of France, Germany and the United Kingdom the ILL provides an extension to the facilities of some 400 university and other laboratories and the associated prestigious nominal roll of nearly 2000 scientists who compete for the available beam time is without equal. There appears to be no way that government accountants can but regard this as a hyperefficient way to dispense limited national research money to the greatest possible benefit.

Innovation was a constant byword; a hundred young (and recently young) scientists ensured there was no shortage of ideas. Some ideas, such as the hedgehog, a preliminary excursion into the field of multidetectors and now in retirement at the South Kensington Science Museum, statistical choppers and IN9, a polarised proton filter spectrometer, had to be tried even though they were to be eventually abandoned. Others, such as neutron spin echo, supermirrors and the world's most beautiful focusing monochromators added a completely new dimension to the Institute's instrumentation possibilities. The ILL became unique in what it had to offer. Operational efficiences due to specimen flux increases, simultaneous shaft settings and eventually multidetectors were soon to signal the demise of Nicole and Carine, the original control computers, in favour of dedicated versions where on-line preliminary data analysis and fast response times became supplementary experimental features put at the disposal of visitors.

From twenty experiments to over 800 per year, was the productivity increase the Scientific Secretariat was presented with and no precedent for organising some thousands of active scientists from many countries, so that they arrived at the ILL at a precise time and date. Concessions were inadmissible. We were expected to know that certain samples were unstable and could only be prepared at the last minute, that lecturing and other duties placed severe constraints on availability and that school holidays of the offspring were not, least of all, an insignificant factor in the forward planning. Neutron production is an expensive business and a motto for each instrument could well have been 'Thou shalt not stand idle.'

To understand completely this self-applied maxim one would have to uncover little-known statistics and admit that of course things went wrong; of course things became chaotic. For example the ILL reactor, one of the most advanced in the world, is only 98% reliable and in one of our darkest moments in 1983, even fell to 97.1% in its availability to provide neutrons for in-house staff and visitors. Two per cent is negligible; unless of course you are in an aeroplane flying from Berlin or Edinburgh to keep a long planned rendezvous with a triple-axis or time-of-flight spectrometer and are yet unaware that a thunderstorm has tripped the reactor leaving you with a 48 h poison-out period to spend in down-town Grenoble. Was the

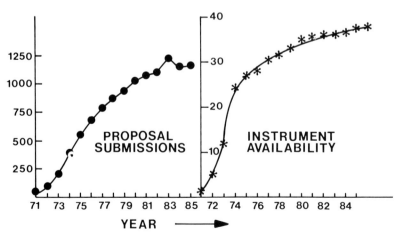

Figure 5.13 The increase in the number of proposals submitted each year, from 1971–85, together with the growth in the number of instruments available for use.

conversation at dinner the previous evening *really* challenging the necessity of these frequent, all expenses paid, trips to the South of France?

Hurried schedule changes, urgent telephone calls and above all a friendly first name approach were the essential ingredients necessary to re-establish the programme in this quite esoteric business. Practice and long experience in public relations had always been necessary due to the massive over-subscription in the bi-annual competition for beam time. From 650 submitted proposals some 300 were always destined for rejection as even the ILL facilities have a limit. These were often excellent proposals too, and words had to be found to make a negative reply sound encouraging—'It was not accepted due to heavy overload on the requested instrument.'

Production line science, PR and publicity, customer profile assessment! Strange words and understandably foreign to a scientist in the Ivy League but, in the run-up to the 21st century, attitudes are, of necessity, changing rapidly. Efficient business methods, tempered of course with a personal approach, are being accepted as essential in our organisational efforts, uniquely directed towards assisting individuals with their scientific problems. Background criticism is always present, but in the final analysis, the superb quality of the ILL facilities may mute all but the most hardened conventionalists.

At the outset, this organisational problem was confided to two people, not forgetting, of course, the same number of indefatigable secretaries. It was obvious though, that with an increase in the number of visitors of between one and two orders of magnitude, staffing problems in the secretariat (which also had many other pseudo-scientific tasks over and above the user programme) would become severe. This was finally solved and by 1985 the office configuration was . . . 2 people, not forgetting of course the same number of, by now, even more indefatigable secretaries. Neutrons, it seems, are not the only things considered expensive by the ILL.

Even in the very early days when France and Germany were the only partners, English was accepted as the language for scientific communication. With the rapid expansion of the scientific programme some form of ILL bureaucracy, which for 70% of the visitors was in this foreign English language, had to be invented. It had to be carefully invented, for not only is bureaucracy anathema to the majority of scientists, but also it was essential that proposal forms, report forms, invitation letters etc had to be unambiguous both in Peking and Bruyères-le-Châtel. However, we succeeded to a certain extent for it is to our satisfaction that we have seen our work copied in other countries.

'Other countries' is a phrase often used elliptically to avoid defining some 30 countries who also participate in the ILL scientific life, alongside the 'members'. It is one of the most significant manifestations of the scientific strength of the ILL that it can say to a scientist, no matter to where he is affiliated, 'this is a good experiment, come and do it at our Institute'. Invariably he will be asked to give a seminar in parallel with his experiment and contribute to the established discussion forum for which the ILL has become renowned, but it is the spirit of international cooperation in the field of pure and fundamental research which has been the real spin-off from relatively moderate governmental financial contributions.

What of the future? The ILL has recently completed a five-year modernisation programme culminating with the installation of new spectrometers and diffractometers benefiting from recent rapid technological advances and with provision for a third generation suite of high-resolution instruments working in the long wavelength part of the spectrum and fed from an additional cold source. This modernisation also affected the reactor, with the replacement of all the beam tubes and the original vertical cold source, prolonging its availability well into the 21st century. Exciting possibilities are also raised with the probable implantation of the European Sychrotron Radiation Source at Grenoble and the benefits which would accrue from being able to do both X-ray and neutron scattering research on the same site. The interval between now and the end of the present century will see the advent and fulfilment of powerful pulsed neutron spallation sources and at these laboratories we wish our scientific colleagues well in their task. Our mentors, the pioneers of neutron scattering in the early 1950s, may well rest assured that their painstaking research, often under very primitive conditions, has been passed on to a succeeding generation of scientists no less enthusiastic and excited by this powerful technique.

5.3.3 I began as a Scientific Assistant: Harwell to Grenoble via Lucas Heights

D A Wheeler, Institut Laue–Langevin, Grenoble, France

Since my very early schooldays my intention had always been to learn to fly with the Royal Air Force, it was only on being rejected following an eye accident that I was obliged to consider an alternative—but what?

It was 1953 and Harwell were advertising for Scientific Assistants in the new field of Atomic Energy Research. I started there at 17 straight from school, with a salary of £18.6.8 per month, and spent the first 6 weeks at the training school learning the principles of vacuum techniques, electrical wiring, glass-blowing, and the use of machine tools.

As a fully-trained Scientific Assistant I was then allocated to the Neutron Diffraction Group (there was no choice offered or any evident selection procedure), with a first task of calculating by hand the structure factors of potassium dihydrogen phosphate for Bas Pease. He was soon to move to the Zeta project, but I opted to stay in the Diffraction Group and work for Ray Lowde, a decision which was to influence completely the rest of my life. I learnt from him the necessity for preciseness, accuracy, attention to detail, and to be prepared for the unexpected. It was vital to be tall and athletic to work with Ray, as our two spectrometers on BEPO were both more than 6 feet from the floor, and one of them could only be adjusted by standing astride an open tank of varying organic liquids used for criticality measurements by the reactor physicists.

The reactor was shut down every Monday morning for fuel changes, which allowed for collimator changes, alignment and modifications, and we regularly had to 'harangue' the shift operators to allow us more time to finish a critical measurement before they started up. Lining up of instruments was straightforward, a light at one end of the reactor hole, and indirect visual sighting through mirrors from the other end. Collimators were made in the carpenter's shop, and the source block position was checked and adjusted with a calibrated rod. Our spectrometers had been constructed and adapted on site—including the detectors. Angular measurements were direct-reading with a scale and a pointer, a simple vernier, or, for the vertical movements, by the use of machined steel spacer blocks inserted under the counter shield which was suspended in a pantograph. To adjust the height required two persons, one of whom had to hang from a not too conveniently located platform to take the weight of the counter shield.

Counting periods were determined by stop-watch, and the entire working day was spent in manually changing temperatures, stepping, timing, counting and plotting Bragg peak intensities. When results were needed for a conference, there was nothing for it but to work longer hours, and the technique was to use a large car parked just outside the reactor, and to take turns in sleeping and working day and night. When doing temperature scans, in order to get a good start to the day, ice cubes for the cold junction were brought to work in a thermos flask, as we could not always be guaranteed access to the only refrigerator in the building, which was in the tea room. No monitor counters were available, and regular visits to the control room were necessary to check on the control rod positions, so as to calibrate our results later. The stability of the counting electronics was never fully trusted, and complete calibration of the electronics and bias curves were carried out weekly.

The radiation levels were high and the weekly notifications of overdose were used to decorate the laboratory—a laboratory which served for up to six of us, packed with desks, lathe, drilling machine, welding bench and drawing board.

Encouragement was given to all scientific staff to pursue further studies, by allowing one day off per week to follow Ordinary and Higher National Certificates in Applied Physics, which I did at Reading Technical College. At the age of about 22 I became an Assistant Experimental Officer.

During this early period DIDO and PLUTO were being constructed and more sophisticated equipment and techniques were needed: we were even able to progress from a bicycle to a van for getting around the site. Collimators had now to be in steel, with rotating shutters, and following a magnificent display in which we set fire to all imaginable variations of shielding materials, the old BEPO paraffin-filled wooden shielding boxes were condemned as no longer acceptable in a sealed building—this was a great shame as they had so many other uses as seats, tables, steps, and as height adaptors for experimental equipment. We then turned to polythene scraps and resin for counter shields and sheets of impregnated compressed wood for primary shields.

Development of high-speed rotor materials was carried out, and experience gained in spinning, balancing, damping of oscillations, and phasing multi-chopper systems. Many materials were tested to destruction, including the newly available resin-impregnated glass fibres, occasionally with catastrophic results as on one memorable occasion when a burst rotor escaped from its test tank. We also branched out into trying to grow our own single-crystal samples and monochromators.

More automatic control was coming in but confidence in leaving a run for the weekend was still limited, and most weekends required visits to change the sample or check the run. It was not until the early 1960s that the development of a fully automatic-controlled neutron single-crystal diffractometer was carried out with Ferranti and Hilger, following the development of the X-ray model. This was an interesting project with regular periods spent in Edinburgh, and provided a first opportunity to fly in an airliner and to be introduced to real whisky.

There were always regular visiting scientists on attachment, and complete freedom of exchange of ideas and techniques. Among them, Harold Smith from Oak Ridge being the first to introduce us to the Polaroid neutron camera, Max Hatherly from Sydney who taught us the techniques of alloy sample preparation and heat treatments, as well as Australian sandwich lunches, Hirakawa from Japan, Umakantha from India and Sabine from Lucas Heights who returned home with several naval gun mounts, which I was later to come to know well.

It had been the norm during the expansion of the Atomic Energy Authority to expect to be promoted to Experimental Officer at about the age of 26, but by 1964 this was no longer the case, and to get my promotion I left Neutron Scattering and joined the Science Research Council, at Rutherford Laboratory. This move almost immediately produced an invitation to return to Neutron Scattering at Lucas Heights, Sydney, working with the Australian Institute of Nuclear Science and Engineering (AINSE).

My interview consisted of a discussion over an evening meal at the Randolph Hotel in Oxford, with Bill Palmer the Scientific Secretary, and by the time the coffee arrived I had a hand-written contract of employment as a Research Assistant. Such simple administrative procedures were a complete change from that of the Authority and the SRC, but on the basis of this flimsy contract we sold our home, packed up our furniture and left for the Antipodes on a five-week boat trip. My task was to develop the necessary neutron scattering diffractometers and equipment for use by the Australian Universities at the Australian Atomic Energy Commission (AAEC) reactor, HIFAR, in Sydney. In this way, AINSE became the first in the world to have an organised nationwide university user programme.

The change in lifestyle was abrupt, from a maisonette, a bicycle and the bus to work, we were able to live in a house, have a telephone, and buy a car. The working and social atmosphere was completely different, the existence of car-pools for travel to work ensured regular contact between technicians and scientists from completely different disciplines; lunchtime football, cricket and chess competitions also served to break down barriers between departments. Coffee mornings were organised to welcome the newcomers, and parties were a regular feature of life.

The HIFAR reactor and principal buildings were identical to DIDO, and it was therefore uncanny to be in a familiar environment surrounded by similar equipment and yet be 12 000 miles from home.

The AINSE inherited from the AAEC a powder instrument based on one of the naval gun mounts from Britain (figure 5.14), a single-crystal instrument based on a converted GEC X-ray spectrometer, and a non-functioning helium bath cryostat. We were fortunate to recruit almost immediately a technician, Peter Lloyd, with an amazing capacity to tackle almost anything, to talk his way out of any situation, with a common fervour to get things done quickly, and who always had an eye for a bargain.

It was an exciting period of development with students coming from all over Australia and others from India, Pakistan, and Malaysia, encouraging us to develop equipment and techniques. With them we carried out the first neutron scattering experiments at helium temperatures in the Southern Hemisphere, using the repaired bath cryostat, but soon had a variable temperature cryostat operating for temperatures from 2 K and in magnetic fields up to 10 kOe. To get the liquid helium involved a 50 mile round trip to fill our only 25 litre dewar, for a price five times that current in

Britain. Needless to say our technique of transferring the liquid and changing of samples was refined to the maximum to minimise losses. Careful planning of experiments was essential, especially as by a slight detour on a Friday we could collect cheap frozen chickens for weekend barbecues.

The development of the instruments and the Institute continued, a scientist and a second technician were recruited, and a water-cooled source block was installed in a tangential hole, giving us two high intensity single-crystal diffractometers on one end and a small-angle instrument on the other end. With the collaboration of the AAEC, one of these instruments was developed into a fully automatic one, using a PDP-8 computer. A second powder instrument was also constructed, as well as a computer-controlled polarised neutron diffractometer with tilting counter, and an instrument for long-wavelength polarised neutron measurements. Another development was a miniature device for working at liquid nitrogen temperature on a single-crystal diffractometer. It was based on the JT expansion of nitrogen gas, using a cryotip (figure 5.15) which had been developed for space research. Although a number of experiments were carried out, we had great difficulties with the purity and dryness of the gas.

The disadvantages of working in Australia were evidently that overseas visitors were rare, equipment took a long time to be delivered, and contact with manufacturers and suppliers was not easy, but by judicious use of university workshops and bending of rules much was achieved. From a personal point of view the advantages of the weather, the proximity of the sea, the National Parks, and yet only being 20 miles from Sydney, more than compensated for any disadvantages.

In 1973 Britain joined the Institut Laue–Langevin, and following an application I received a telegram inviting me 'to come to Grenoble for one day next week for interview', I was convinced that the Administration thought I lived in Austria and not Australia. With the help of the bank I duly arrived in Grenoble via Hong Kong, but in common with many others after me, it was somewhat of a shock to find how difficult it was to make oneself understood, and my first feelings were not encouraging. But the desire to participate in a newly developing organisation was too much of a challenge not to accept the offer of an engineer post as an Assistant to the Head of the Reactor Department, so in 1974 we sold our house, packed our furniture, and took the now standard 24-hour route to Europe by air.

My present position is as Deputy to the Head of the Instrument Operation Department with a staff of 12 engineers and 82 technicians, who are responsible for the day-to-day running of the instruments and the assistance to the 76 ILL scientific staff and some 1500 scientific visitors per year. In addition we have the responsibility for the maintenance and construction of the site buildings and services. My job now involves a great

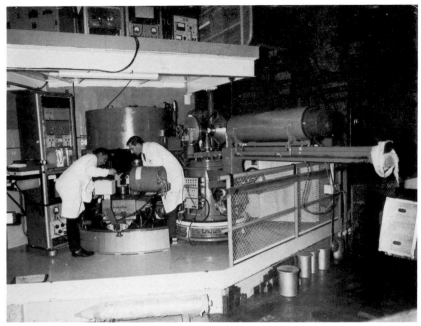

Figure 5.14 Frank Moore and David Wheeler working at the 4H1 powder diffractometer, based on a naval gun mount, at Lucas Heights, Australia.

deal of administration and budgetary questions, personnel problems and
organisation, the major part conducted in French.

After 11 years in France and looking back over more than 30 years in the
neutron scattering field what has changed?

We have a more powerful source of neutrons, more experimental
positions thanks to guide tubes, focusing monochromators and multi-
counters, automatic angular read-out and stepping motors, computer-
controlled diffractometers, direct graphical plots of the results and
telephone links from home to the computers. There is no need for
today's technician to spend all day counting, stepping and plotting; quite
the reverse for in many cases he is frequently preparing the experimental
configuration for the next visitor. The working hours are shorter, the
holidays longer, the salaries and working conditions much better.

The safety around the instruments has certainly evolved for the better,
with an increase in alarms, interlocks and automatic shutting of beam lines.
More attention has been given to radiation background levels, not only for
biological reasons (the acceptable limits for a working area have been
reduced by a factor of three) but to reduce unwanted counts in the
detectors. Proper working procedures are now clearly defined and interlocks prevent accidental irradiation. Whilst lead bricks have remained the

Figure 5.15 A computer-controlled 2-circle diffractometer with a
cryotip cooling device at Lucas Heights, 1970.

most practical γ shielding, the development of flexible boron carbide
sheets, sintered lithium fluoride, polythene and gadolinium paint have
changed the form of neutron shields. More stringent rules are applied
before authorisation is given for any experiment, mainly due to the types of
samples now being studied and the increasing use of continuous-flow
explosive gases for detectors. The sophistication of the sample environment with extremes of temperature and pressure, and high field intensity
superconducting magnets, has led to the sample environment becoming a
specialised sector of its own, involving 18 persons full-time at the ILL.
Liquid helium is now delivered daily in 100 litre dewars and consumed by
the thousands of litres per month.

However, despite all these improvements my impression is that the
majority of technical staff now have a much lower involvement in the
actual experimental measurements than 30 years ago, and that there has
also been a marked tendency to reduce the responsibility and role of the
'scientific assistant'.

Since my time as a Scientific Assistant at Harwell, I have achieved my
aim to be a pilot, by learning to fly sailplanes in the Alps, have had the

opportunity to meet people of many different nationalities, to work for many distinguished scientists, and to help many students to obtain their doctorates, but there are still a number of years before retirement and my training taught me that one should always be ready for the unexpected.

6 Some Statistics from the Neutron Diffraction Commission

B T M Willis

Chemical Crystallography Laboratory, University of Oxford and Atomic Energy Research Establishment, Harwell

Activities of the Neutron Diffraction Commission, 1969–85

At the Sixth General Assembly of the International Union of Crystallography, held in Rome in 1963, it was proposed that the Union should consider the formation of a Commission on Neutron Diffraction, and Professor G E Bacon was invited by the Executive Committee to study the proposal and to report back before the Seventh General Assembly in 1966.

In his report to the Executive Committee in 1965, Professor Bacon expressed the general feeling of workers in the field of neutron diffraction that there were a number of essential functions which a new Commission could perform. These functions included:

(*a*) Tabulation and critical evaluation of data on the neutron scattering amplitudes of elements and isotopes (including complex scattering amplitudes).

(*b*) Tabulation of magnetic form-factors.

(*c*) Recommendations on technical procedures for carrying out diffraction experiments, e.g. choice of collimation; choice of monochromators; and corrections for secondary extinction and multiple scattering.

(*d*) Cataloguing of information on reactor types, neutron fluxes, instruments, and methods of data-collection and analysis.

(*e*) Encouragement of monographs on specific aspects of neutron scattering; cooperation with the editors of *Structure Reports* (published for the

International Union of Crystallography) in the description of magnetic structures.

(*f*) Support for symposia, to ensure that the various branches of neutron diffraction are covered adequately at the triennial meetings of the IUCr.

Formal approval for the setting up of a Commission was given in Moscow at the 1966 General Assembly. Professor Bacon became the first chairman and there were six elected members (from Norway, France, Canada, USA and USSR). The names of the chairmen of the Commission since its inception in 1966 are:

G E Bacon (UK)	1966–9
L M Corliss (USA)	1969–75
A F Andresen (Norway)	1975–8
D E Cox (USA)	1978–81
M S Lehmann (France)	1981–4
B T M Willis (UK)	1984–7

In the following paragraphs we describe briefly the progress in meeting the objectives under (*a*) to (*f*).

(*a*) *Neutron scattering amplitudes*
Between 1966 and 1969 data were assembled on the neutron scattering amplitudes of the elements and isotopes, and a table was then published in *Acta Crystallogr.* (1969) **A 25**, 391. This table contained forty additional values and thirty modified values compared with those listed in the *International Tables* **II**, 227 (1962). An improved compilation was published several years later in *Acta Crystallogr.* (1972) **A 28**, 357. This, in turn, has now been superseded by the table prepared by Koester†.

(*b*) *Magnetic form-factors and magnetic structures*
A compilation of magnetic form factors has been carried out over several years by W C Koehler and R M Moon of the Oak Ridge National Laboratory, USA. In 1979 the work was continued as a joint project with the Commission on Charge, Spin and Magnetization Densities. Subsequently, the compilation of Koehler and Moon was taken over, on behalf of the Neutron Diffraction Commission, by J X Boucherle of Grenoble, France. Boucherle incorporated data from various sources, including a list prepared by A Delapalme for the Seventh Sagamore Conference (Japan 1982), and created a computer file of Magnetic Form Factors and Magnetization Densities. Any modifications can be easily inserted in this file, whose first version was made available in March 1984.

In 1969 the Commission established an information service for the rapid dissemination of new results on magnetic structures. This took the

† See Koester and Steyerl (1977).

form of Magnetic Structure Data Sheets, which were distributed quarterly to subscribers (and financed through a modest subscription fee). The data sheets were prepared for insertion into a loose-leaf binder, similar to the first edition of Wyckoff's 'Crystal Structures'. The MSDS file was handled for ten years by D E Cox of Brookhaven National Laboratory, USA, and was transferred in 1982 to S Murthy of Trombay, India. The total number of entries in the MSDS file exceeded 300 in 1985.

(c) *Recommendations on technical procedures*
We refer to this in the next section.

(d) *Cataloguing of information*
A newsletter was introduced in 1974, edited by B Klar of Grenoble. The editorship was taken over by W B Yelon of Missouri in 1977. From 1984 a new arrangement was adopted, whereby the editor (a member of the Neutron Diffraction Commission) was changed after each issue. The advantage of this scheme was that the Commission membership reflected world-wide interests, and the editor could concentrate on that part of the world he knew best. Two issues of the newsletter go out each year to over 800 people.

(e) *Encouragement of monographs*
This is one area where progress has been disappointing. Compared with the field of X-ray diffraction, there are relatively few monographs on neutron diffraction, and more are needed.

(f) *Support for symposia*
The Commission has organised a number of satellite meetings on neutron diffraction, held immediately before or immediately after the main triennial Congresses of the IUCr. These meetings were held at Stony Brook (1969), Petten (1975), Cracow (1978), Argonne (1981) and Berlin (1984). None of the Proceedings was published, with the exception of the meeting at Petten. (Report of Reactor Centrum, Nederland, CRN–234 (1975).)

Intercomparison Project of the Commission

In 1969 the Commission initiated a project for the intercomparison of powder-diffraction instruments operating at different reactor centres. Its purpose was to help experimentalists to obtain the optimum performance from their instruments. The project was organised by a small group consisting of A F Andresen (Norway, Chairman), T M Sabine (Australia) and R P Ozerov (USSR). A similar survey had already been carried out for South East Asia by Sabine.

 In contrast to X-ray equipment, instruments for neutron diffraction are individually built, and so their performance can be expected to vary considerably from one instrument to another. However, it was considered that powder-diffraction instruments were sufficiently uniform in design to be used for the intercomparison. One tries with such instruments to obtain powder patterns with high intensity, good resolution and a large peak-to-background ratio. Clearly, good resolution can only be achieved by sacrificing intensity; and both resolution and intensity can be enhanced by increasing the wavelength, but only at the expense of limiting the number of observable reflections.

The participants in the Intercomparison Project were asked to do two things: irradiate gold foils placed in the path of the monochromatic beam, so that the intensity of the incident beam and its rate of fall off with distance could be determined; and record the powder diffraction pattern of a standard Al_2O_3 sample, from which the intensity of the diffracted beam and the instrumental resolution could be extracted. Standardised gold foils and Al_2O_3 pellets were distributed to 31 laboratories, together with detailed instructions as to how to proceed with the measurements. Complete sets of data were obtained from 21 laboratories over a time span from 1971 to 1974. The final results of the intercomparison were published by Andresen and Sabine (1977).

One of these results is illustrated in figure 6.1, where the quantity $\Delta d/d$ is plotted against $S \cot \theta_M$. d is the interplanar spacing for a reflection and

$$\frac{\Delta d}{d} = \tfrac{1}{2} A_{1/2} \cot \theta_B$$

where $A_{1/2}$ is the FWHM of a peak with a Bragg angle θ_B, while S is the opening angle of the first collimator and $2\theta_M$ the take-off angle of the monochromator. Good resolution, as indicated by small values of $\Delta d/d$, is

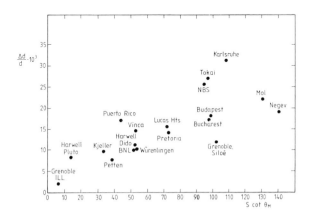

Figure 6.1 The dependence of angular resolution on collimator angle and take-off angle at the monochromator, for powder diffractometers at various Institutes.

expected for small values of the function $S \cot \theta_M$. This is the general trend of the data in figure 6.1. The instrument D1A at the High Flux Beam Reactor in Grenoble has by far the best resolution, using a very high take-off angle ($2\theta_M = 122°$). The performance of D1A is described by Hewat (1976); this is widely regarded as the prototype instrument of the next generation of reactor-based powder diffractometers.

Survey of Neutron Sources for Diffraction Work

In this section we give some statistics of research centres at which experiments in neutron diffraction are carried out. Most of this work is carried out using nuclear reactors as neutron sources, but an increasing number of centres are using accelerator-driven neutron sources.

Reactor neutron sources

These are listed in table 6.1, which is not comprehensive, but includes those institutes responsible for more than 80% of the publications in neutron diffraction.

The last column in the table is the peak thermal flux, as measured in the core near the end of the beam tubes. The reactors with the highest flux (1.2×10^{15} neutrons cm^{-2} s^{-1}) are the HFIR at Oak Ridge which became critical in 1965, and the HFR at the Institut Laue–Langevin, Grenoble, which was first critical in 1971. There has been little further progress in improving the neutron flux from reactors since 1965.

The table shows that very few research reactors were built in the 1970s and early 1980s. However, this lean spell is now being followed by a period in which there will be a major world-wide re-investment in research reactors, either as major upgrades of existing reactors or as completely new facilities.

Table 6.1 Reactor neutron sources

Country	Institute	Name of reactor (with year of start up)	Reactor power (MW)	Peak thermal flux (neutrons $cm^{-2}s^{-1}$ $\times 10^{14}$)
Australia	AAEC Lucas Heights, New South Wales	HIFAR (1962)	10	1
Austria	Atominstitut der Osterreichischen Universitaten, Vienna	TRIGA (1962)	0.25	0.1
	Austrian Institute for Atomic Energy, Seibersdorf	ASTRA	8	1.3
Belgium	CEN/SCK, Mol	BR2 (1963)	55	5
Canada	Chalk River Nuclear Laboratories, Ontario	NRU (1957)	135	3
	Nuclear Research Centre, McMaster University, Ontario	McMaster University Reactor (1959)	2	0.3
China	Institute of Nuclear Energy, Beijing	HWRR (1958)	15	2.8
Denmark	Risø Research Establishment, Roskilde	DR-3 (1960)	10	1.5
Finland	Technical Research Centre of Finland, Espoo	FIR-I (1962)	0.25	0.01
France	Institut Laue–Langevin, Grenoble	HFR (1971)	57	12
	Laboratoire Leon Brillouin, Gif-sur-Yvette	ORPHEE (1980)	14	2
	Centre de l'Energie Nucleaire, Grenoble	Melusine	8	—
East Germany	Zentral Institut fur Kernforschung, Rossendorf, Dresden	RFR (1957)	10	0.1

continued

Table 6.1 *continued*

Country	Institute	Name of reactor (with year of start up)	Reactor power (MW)	Peak thermal flux (neutrons $cm^{-2}s^{-1}$ $\times 10^{14}$)
West Germany	Hahn–Meitner Institut, Berlin	BER-II	5	0.2
	GKSS Forschungs-zentrum, Geesthacht, Hamburg	FRG-I	5	0.3
	Institut fur Fest-korperforschung, Julich	FRJ-2 (1962)	23	2
	Reaktorstation, University of Munchen, Garching	FRM (1957)	4	0.2
Hungary	Central Research Institute for Physics, Budapest	WWR-SM (1959)	4	0.9
India	Bhabha Atomic Research Centre, Trombay, Bombay	CIRUS (1960)	40	0.6
		DHRUVA (1984)	100	1.8
Indonesia	Research Institute for Nuclear Techniques, Bandung, Java	TRIGA-II	1	0.5
	Centre for Materials Testing, Serpong, Jakarta	MPR-30 (1987)	30	2
Israel	Nuclear Research Centre, Negev, Beer-Sheva	IRR-2 (1964)	26	0.5
Italy	ENEA/CRE, Casaccia, Rome	TRIGA-II (1961)	1	0.2
Japan	Japanese Atomic Energy Research Institute, Tokai, Ibaraki	JRR-2 (1960)	10	1
	Research Reactor Institute, Kyoto University, Osaka	KUR (1961)	5	0.3
Korea	KAERI, Seoul	TRIGA-III	2	0.8
Malaysia	UTN, Puspati	TRIGA-II (1982)	—	0.1

continued

Table 6.1 *continued*

Country	Institute	Name of reactor (with year of start up)	Reactor power (MW)	Peak thermal flux (neutrons $cm^{-2} s^{-1}$ $\times 10^{14}$)
The Nether-lands	Joint Research Centre, Petten	HFR (1961)	45	3
	Interuniversity Reactor Institute, Delft	HPR (1963)	2	0.2
Norway	Institute for Energy Technology, Kjeller	JEEP-II (1966)	2	0.3
Pakistan	Pakistan Institute of Nuclear Science and Technology (Pinstech) Islamabad	PARR (1965)	5	0.1
Philippines	PARC, Quezon City	PRR-1 (1962)	1	0.1
Poland	Institute of Atomic Energy, Swierk, Otwock	EWA (1958)	10	1
		MARIA (1974)	30	2
Puerto Rico	Puerto Rico Nuclear Centre	TRIGA	1.2	0.2
Romania	Institute of Physics and Nuclear Engineering, Bucharest	TRIGA (1957)	2	0.2
South Africa	Pretoria Nuclear Research Centre	ORRR	20	2
Sweden	Studsvik Energiteknik AB, Nykoping	R2 (1960)	50	2
Switzerland	Eidgenossische Institut fur Reaktorforschung, Wurenlingen	SAPHIR (1958)	10	1.4
Thailand	Office of Atomic Energy for Peace	TRIGA-II (1965)	—	0.1
UK	Atomic Energy Research Estab-lishment, Harwell	DIDO (1965)	25	2
		PLUTO (1957)	25	2
USA	Brookhaven National Laboratory, Upton, New York	HFBR (1965)	60	10

continued

Table 6.1 *continued*

Country	Institute	Name of reactor (with year of start up)	Reactor power (MW)	Peak thermal flux (neutrons $cm^{-2}s^{-1}$ $\times 10^{14}$)
USA	Massachusetts Institute of Technology, Cambridge	MIT Research Reactor (1958)	5	1
	Research Reactor Facility, University of Missouri, Columbia	MURR (1966)	10	1.2
	NBS Research Reactor, National Bureau of Standards, Washington	NBSR (1967)	10	2
	Oak Ridge National Laboratory, Tennessee	HFIR (1965)	100	12
USSR	Kurchatov Institute of Atomic Energy, Moscow	RR-8 (1981)	8	2.5
	Laboratory of Neutron Physics, Joint Institute of Nuclear Research, Dubna	IBR-2 (1979)	4	—
Yugoslavia	'Boris Kidric' Institute of Nuclear Sciences, Vinca	TVRS	6	0.6

Accelerator-driven neutron sources

No survey of neutron diffraction instruments and institutes would be complete without referring to the new generation of accelerator-driven neutron sources. In 1986 there were four proton spallation sources operating in the world: at the KEK Laboratory (Japan), at Los Alamos National Laboratory (USA), at Argonne National Laboratory (USA) and

Table 6.2 Accelerator driven neutron sources.

Country	Institute	Source name (with starting date)	Particle energy (MeV)	Time average current (µA)
Japan	KEK Tsukuba	KENS-I (1980)	500 (protons)	2
		KENS-I (1985)	500 (protons)	10
UK	Rutherford Appleton Laboratory	ISIS (1985)	800 (protons)	200
	Atomic Energy Research Establishment, Harwell	HELIOS (1980)	136 (electrons)	100
USA	Los Alamos	LANSCE (1985)	800 (protons)	100
	Argonne National Laboratory	IPNS-I (1981)	500 (protons)	12

at the Rutherford Appleton Laboratory (UK). Table 6.2 presents a few statistics of these pulsed neutron sources.

The development of pulsed reactors for experiments in pulsed neutron scattering belongs to the USSR. A series of pulsed reactors has been operated at the Joint Institute for Nuclear Research at Dubna. The first pulsed reactor, the original IBR, began operating in 1960. The latest pulsed reactor, IBR-2, became critical in 1979. Pulses of short duration, about $3\,\mu s$, are achieved in IBR-2 by running it in its booster mode.

Pulsed sources open up a number of new possibilities, for example, experiments with epithermal neutrons (energies between 0.1 and 10 eV), and experiments in extreme environments. There is every prospect of exciting work in neutron diffraction in the 1990s and beyond!

Acknowledgments

In compiling table 6.1 the author is most grateful for assistance from the following: A Albinati, A F Andresen, D Bally, N M Butt, R Chidambaram, D E Cox, G Dolling, K Hennig, A W Hewat, P Hiismaki, J Leciejewicz, I Olovsson, W Prandl and H G Smith.

References

Andresen A F and Sabine T M 1977 *J. Appl. Crystallogr.* **10** 497
Hewat A W and Bailey I 1976 *Nucl. Instrum. Methods* **137** 463
Koester L and Steyerl A 1977 *Springer Tracts in Modern Physics* vol 80 (Heidelberg: Springer)

7 Some Crucial Applications

7.1 Neutrons and Magnetism—a Brief History

W C Koehler

Oak Ridge National Laboratory

Introduction

Magnetism in matter is a discipline that has attracted the attention of scientists and engineers for a very long time. It continues to attract their attention because of its importance in understanding the structure of matter on a microscopic scale, and to the evolution of technologies that depend to a greater or lesser extent on such microscopic structures. The study of magnetism near phase transitions has provided an arena particularly conducive to the productive confrontation of theory with experiment. Of the many probes that are used today to investigate magnetism in matter, none is so ideally suited for the purpose as the neutron†.

Thermal neutrons, those brought into thermal equilibrium with a moderator at or near room temperature by successive collisions, have properties that make them unique for the study of magnetic structures and dynamics, namely, wavelengths comparable to interatomic separations, energies of the same order of magnitude as those of elementary excitations, zero charge and an intrinsic magnetic moment. Neutrons undergo interactions with the electronic moments of atoms in matter that result in scattering. In the scattering cross section, a measure of the efficiency of scattering, is contained the information that one hopes to obtain about the scatterer. This cross section is cast in various forms chosen according to the types of questions that are posed by the experiment. These questions have changed over the past nearly half a century following advances in the

† B N Brockhouse has remarked that 'If the neutron did not exist, it would need to be invented'.

strengths of neutron sources, developments in theory, and evolution of scientific knowledge generally. The ingenuity of the questioners in developing new instrumentation has played an exceptionally important role.

With a few exceptions neutron scattering experiments are concerned with ordered systems of spins (moments), in particular with their static configurations, and with the motions of the spin systems at various temperatures relative to the ordering temperature. These spin configurations and their motions are determined primarily by exchange forces and the experiments can potentially reveal the strength and range of such forces and how these quantities vary with temperature. Perturbations on the exchange energy such as those due to crystal field, magnetoelastic, and magnetic dipole interactions, and to an applied magnetic field can sometimes also be extracted from the experiments.

The ordered configurations of magnetic flux quanta in certain classes of superconductors and their interactions with defects and impurities, the very interesting competition between magnetic ordering and superconductivity in the class of substances known as re-entrant superconductors, and the nature of processes that produce heavy fermion superconductivity are also amenable to study by magnetic neutron scattering.

In this chapter I have tried to trace the evolution of magnetic neutron scattering and I have chosen to correlate it with the development of research reactors as sources of neutrons. This does not mean that I have overlooked the importance, sometimes crucial, of accelerators as neutron sources, and in the future, another history written by another author, may be correlated with the development of pulsed neutron sources. Yet, it is fair to say that up to now experiments at reactors have been more important for the increase in our knowledge of magnetism in matter than have those at accelerators.

Finally, this article is not intended to be a complete review of scientific achievement. It is as the title indicates a brief recounting of events, in some of which I was an active participant; of others, I was a contemporary observer.

Prereactor Period 1932–42

Neutron physics had its origin in 1932, the year that marked the discovery of the neutron by Chadwick. The fiftieth anniversary of that event was celebrated by a large conference held in Cambridge in 1982, at which all aspects of neutron science and politics were discussed (Schofield 1982). Some repetition of observations made there may be inevitable here.

Neutron scattering as applied to condensed matter science may be said to have started in 1936 with the demonstration of the phenomenon of Bragg diffraction. Three papers on this matter are reprinted earlier in this volume. The sources used were called neutron howitzers. They consisted of a Ra–Be mixture surrounded by moderating and shielding material, and they were weak and polychromatic. Even so, this early prereactor period was notable for intense activity in the area of the magnetic scattering of neutrons, much of which was stimulated by a suggestion by Bloch (1936). In 1936, there was good theoretical reason to believe that the neutron had a magnetic moment of the order of one or two nuclear magnetons. Bloch proposed a method for measuring it by means of the interaction of slow neutrons with magnetised ferromagnetic materials and he predicted that polarised beams of neutrons should be produced by transmission through magnetised samples. Shortly thereafter the phenomenon of polarisation was detected, and estimates of the magnitude and sign of the moment were announced (Hofman *et al* 1937)†.

Probably the most fruitful developments in this period were in theoretical investigations. This despite the fact that neither the spin nor the moment of the neutron was known with certainty. There was as well some disagreement, initially, on the form of the magnetic interaction of the neutron with electronic moments. In his original paper Bloch had chosen a form for the interaction that was equivalent to treating the neutron as a magnetic dipole. In a subsequent calculation Schwinger (1937) considered the neutron as an Amperian current with rather different results. In a second paper Bloch (1937) calculated the cross section for scattering from a ferromagnet on both assumptions, showed that there was a potentially easily distinguishable angular dependence to the scattering for the two cases and suggested that it be left to experiment to make the final decision. (At that time, 1937, experiment could not make a distinction because of the unsuitability of the existing sources. Only later, after research reactors came into existence, could the appropriate experiments be done.) Shortly thereafter, in 1938, in a paper that had escaped my notice and evidently that of many others, Midgal (1938) showed that the cross section does not, in fact, depend upon the magnetic nature of the neutron and that the expression obtained by Schwinger is the only correct one.

In any case, that was the point of view adopted by Halpern and Johnson whose classic paper on the magnetic scattering of neutrons appeared in 1939. Their paper was the definitive work on the subject for many years. It contains predictions that were still being verified 30 years after its publication.

The period from 1939 to 1942, the later prereactor period, was notable for the development of the pulsed cyclotron technique for producing neutrons. Such neutrons were used by Alvarez and Bloch (1940) to measure the neutron magnetic moment. They used a resonance technique in combination with an iron polariser and analyser and they obtained a

† See also Powers *et al* (1937, 1938) and von Halban *et al* (1937, 1938).

value close to that predicted from moments measured earlier for the proton and the deuteron. At this time, then, the theoretical foundations of magnetic scattering of neutrons were firmly established.

The Early Reactor Period

The first nuclear reactor was successfully operated in December 1942, under the west stands of the University of Chicago athletic field. In the following year the Oak Ridge Graphite Reactor went into operation, and in 1944, CP-3 (Chicago Pile 3) was started up in a forest preserve south of the city. With these two reactors, one had, for the first time, relatively copious fluxes of thermal neutrons, of the order of 10^{11} to 10^{12} neutrons cm^{-2} s^{-1}, for experimentation. It is around them that the modern phase of magnetic neutron scattering can be said to have begun. Because of intensity limitations (a frequently recurring phrase even today) experimentation was limited to elastic scattering. Diffraction, primarily, was to be emphasised at Oak Ridge and reflection and refraction, primarily, at Argonne. Between 1946 and 1948, the nuclear scattering of thermal neutrons by crystalline powders was systematically investigated by Wollan and Shull (1948a). The first diffractometer was based on an instrument Wollan had used to study X-ray scattering by the inert gases. In the course of measuring the nuclear scattering amplitudes of nearly 100 nuclides, they had to study and untangle the binding, spin, isotope, and thermal effects on scattering as well as to clarify instrumental problems such as multiple scattering in the samples. These questions were rapidly solved and neutron diffraction from powder samples became a well understood and well documented technique (1948b).

The early experiments of Shull and Wollan, up to July 1947, were made by hand. Exposure times were long and it was expedient to make use of several shifts for collecting the data. A Mr M C Marney worked the night shift, and when in due course a system for the automatic recording of data was built and installed it became known as 'automatic Marney'. The first version was exceedingly simple. A microswitch blade falling into a notch on a wheel triggered a pulse to a printing register, a Traffic-Counter, which had recorded the number of scales of 2 (or 4 if a very strong scatterer) accumulated in a given angular interval of the counter motion. Several scans of each pattern were made, the data plotted as a function of scattering angle, by hand, averaged, by hand, and then interpreted. Exposure times were measured in days.

In late 1948 and early 1949 there were undertaken a number of measurements on substances whose atoms possess magnetic moments. Many such compounds had been found to exhibit striking anomalies in the temperature variations of their specific heats, resistivities, and particularly characteristically, their magnetic susceptibilities. They appeared to be undergoing some sort of ordering transition but there was no net magnetisation associated with the supposed ordered phase. An explanation for these phenomena had been proposed quite early by Néel, in terms of an antiferromagnetic configuration, e.g. the atomic moments were supposed to order on interpenetrating sublattices, the directions of one-half of them being opposite to those of the other half. It was recognised by the Oak Ridge group that the Bragg reflections from the magnetic structure would have a characteristic signature: they would occur at positions forbidden to the nuclear reflections. Moreover, they would have, because of the finite extent of the electron distribution responsible for the moment, a form factor, a falling off in intensity with increasing scattering angle. That Néel's hypothesis was correct is of course well known.

Today there exist catalogues of magnetic structures as determined from magnetic neutron diffraction. Some of these structures are quite complicated and could only have been deduced with neutrons. Many structures were solved with powder data, others from single-crystal diffraction after that technique had been put on firm ground by Bacon at Harwell (1951) and by Peterson and Levy (1951) at Oak Ridge.

In the same early studies the correctness of the Néel model for ferrimagnetic (Néel's terminology) Fe_3O_4 (magnetite) was announced (Shull *et al* 1951). The ferromagnetic properties of the ferrites, compounds of great technological interest of which magnetite is the prototype, were shown to arise from incompletely compensated antiferromagnetism. In addition, a rather crude, by today's standards, determination of the magnetic form factor of iron was made from a polycrystalline sample, and reasonable agreement with an early calculation by Steinberger and Wick (1949) was obtained. Studies on single crystals of magnetite laid the foundations for the production and exploitation of polarised neutron beams (Shull 1951). The polarised beam technique would later be used for highly precise measurements of magnetic moment densities with resultant information about the wave functions of those electrons responsible for the magnetic moments.

The Middle Reactor Period

Between 1948 and 1964 there was a great proliferation of research reactors in the world. Many of these were then called high-flux reactors because they had central fluxes of the order of 10^{13}–10^{14} neutrons $cm^{-2} s^{-1}$. Some, to be sure, were of the 10^{12} class, but at all of them began an explosive development of the study of inelastic scattering and slow neutron spectroscopy. In parallel with the new sources, advances in theory and technology contributed to this development.

On the theoretical side, Van Hove (1954a) laid the foundations for what has become an extremely powerful way of treating neutron scattering data with his introduction of the time-dependent pair-correlation function. For magnetic scattering this gives the probability that if at time $t=0$ the magnetic moment at the origin of coordinates $r=0$ has a specified value (magnitude and direction), then the moment at a position r from the origin has some other given value at time t. The importance of this development stems from the fact that important advances in statistical thermodynamics can be applied in certain cases. Now a properly designed neutron scattering experiment measures the Fourier transform in space and time of the correlation function. This Fourier transform, $S(Q,\omega)$, is defined in the momentum–energy (wave vector, frequency) space of the scatterer. It is variously called the scattering function, scattering law, or fluctuation spectrum and it is precisely what is needed to characterise the dynamics of a magnetic system.

In a given physical context there may be other ways to formulate the scattering cross section that are easier to manipulate or are more appropriate to the experimental situation and several other response functions have been defined that link the frequency distribution of scattered neutrons to the fundamental properties of the scattering magnet (Marshall and Lovesey 1971).

On the experimental side several methods, perhaps ten or twelve, were devised to measure inelastic scattering. These differ depending upon the type of experiment envisaged and the range of Q and ω to be observed. For application to magnetic scattering two of these, the triple-axis crystal spectrometer and the chopper with time-of-flight energy analysis, were most important. The first application of crystal spectrometry to a magnetic system was carried out at Chalk River by Brockhouse (1958) in a study of magnetic excitations in magnetite. It was believed then, as it is now, that the thermal disordering at low temperatures of the arrays of ordered moments in ferromagnetic and antiferromagnetic crystals could be described in terms of excitations called spin waves, and by 1957 a considerable literature on the scattering of neutrons by spin waves had come into existence. According to the theory, the scattering is determined by conservation of energy and momentum between the neutron and one spin wave quantum which was called, by Elliott and Lowde (1955), a magnon.

Shortly thereafter, the first dispersion curve, $\omega(q)$, for a ferromagnetic metal, FCC cobalt alloy, was measured on an instrument with which specified directions in reciprocal space could be scanned (Sinclair and Brockhouse 1960). This was the precursor for the very powerful constant Δq and constant ΔE instruments that are ubiquitous today. The measurements were said to be in agreement with the predictions of the Bloch–Heisenberg spin wave theory and a value of the product JS of the exchange integral and the atomic spin could be extracted.

It is appropriate in this discussion of inelastic scattering of neutrons by spin waves that the measurements made by Lowde (1954) on iron and by Riste and his colleagues (1959) on magnetite and haematite be mentioned. These researchers, working at low-flux reactors, used the diffraction method, so called, to make up for the low incident intensity. It had been shown by Elliott and Lowde (1955) that the spin waves manifest themselves as coherent diffuse peaks near the Bragg reflections. In particular, the dispersion relation of the magnons is directly obtainable from the manner in which the width of the diffuse peak changes with the angular mis-setting of the crystal from its Bragg peak. Even though the diffraction method is little used today because of the lack of resolution and detail available, it is a great tribute to these scientists that with physical insight and experimental dexterity they were able to extract so much information with so little intensity at so early a stage.

To conclude this sampling of activities in the Middle Reactor Period, I will mention briefly some of the many investigations that were carried out on elemental iron at temperatures near the Curie point. Latham and Cassels (1952) using cyclotron-produced neutrons and a velocity selector were probably the first to see the effects of magnetic fluctuation scattering on the total cross section of iron near the Curie point. Palevsky and Hughes (1953) made similar observations using the Brookhaven slow chopper. More precise measurements of the sharp increase in total cross section at T_c were made by Squires (1954) with filtered neutrons of average wavelength 7.0 Å.

Following these early experiments that demonstrated 'critical scattering' Van Hove (1954b) formulated an explanation in terms of large scale fluctuations in magnetisation density analogous to the phenomenon of critical opalescence. The critical scattering has two features. Information on one of these, the range of the spin correlations, is obtained from the angular distributions of the scattering as long as the relaxation time of the correlations is much longer than the flight time of the neutron. In this case the neutron 'sees' a static configuration. The earliest studies of critical scattering from Fe to yield a correlation range were measurements of small-angle scattering near the Curie point by Wilkinson and Shull (1956), and Gersch, Shull and Wilkinson (1956) carried out at the Oak Ridge Graphite Reactor.

The second aspect is the relaxation time just mentioned. This is evaluated from measurements of $d^2 \sigma/d\Omega \, dE$. A well defined (in energy) beam is allowed to fall on the scatterer and the broadening in energy is measured as a function of scattering angle. Much of the precise work on inelastic critical scattering in Fe was done by Jacrot and Cribier (1963) and their associates at the EL-3 reactor, 10^{14} neutrons cm^{-2} s^{-1}, at Saclay. They used a neutron velocity selector located at a port looking at a cold moderator. Their measurements were made by time-of-flight analysis of neutrons scattered through various angles. From studies made at 9° scattering angle at the

Curie point a Lorentzian form for the scattering was found, indicating, somewhat surprisingly, that the scattering was inelastic. Because of the significance of these and other observations to the theory of critical scattering an independent measurement of the angular and energy distribution of 4.27 Å neutrons scattered at small angles in Fe near the Curie point was made by Passell et al (1964) and they confirmed the observation made at Saclay that the scattering is not elastic at the Curie point. The story of the implications of these data to the nature of the interactions in Fe is a longer one than I can tell here. A complete discussion, for the time, was given by Marshall and Lowde (1968) a few years later. The saga continues with studies of the excitations at temperatures above the Curie point (Lovesey 1984)†.

The Late Reactor Period

This era extends from 1965 to the present. In the early part of it two high-flux, 10^{15} neutrons $cm^{-2} s^{-1}$, research reactors became operative in the United States, the HFBR (1965) at the Brookhaven National Laboratory, and the HFIR (1966) at the Oak Ridge National Laboratory. These sources were important to scattering studies in general and to investigations in magnetism in particular by reason of their increased intensity. Equally important was the fact that advances in monochromator production, beam tube design, and computer development were all taken into account to produce instruments with state-of-the-art, for that time, capabilities. At Oak Ridge, for instance, each spectrometer was controlled by a PDP-8 minicomputer following a design used by Busing and Levy in driving a 4-circle X-ray diffractometer with a PDP-5. A serious deficiency at these reactors was the lack of capability for flux-tailoring with cold or hot sources. Only recently has a reactor in the United States been equipped with a cold source. In Europe, on the other hand, such installations existed as early as 1956 in BEPO at Harwell and 1959 in EL-3 at Saclay.

Nevertheless, the HFBR and HFIR gave access to data that were up to that time difficult or impossible to acquire. I mention one area of research, among many, that I consider particularly noteworthy at each of these reactors. At BNL great emphasis was placed on critical and spin-wave scattering studies by triple-axis spectrometry. Extremely thorough measurements, with great attention to resolution and absolute intensity considerations, of critical scattering in $RbMnF_3$ (Tucciarone et al 1971) and Fe (Collins et al 1969) were carried out, that put the inelastic magnetic scattering technique on a sound quantitative basis. At ORNL, the fundamentals of what has become known as neutron polarisation analysis

---

† Two very recent papers with different interpretations of the experiments are Shirane (1984) and Mook and Lynn (1984).

were established (Moon *et al* 1969). If in addition to measuring the final momentum and final energy of the scattered neutron, one can ascertain its spin state after scattering, one learns the maximum possible about the scatterer. Polarised neutron scattering is a severely signal-limited technique but it is a very powerful one and will undoubtedly be exploited more fully in future sources.

The early years of this period are remarkable, too, because magnetic neutron scattering matured and consolidated at many of the second generation reactor centres. At Risø, for example, low energy triple-axis spectrometry was emphasised. The first detailed studies of spin-wave scattering in a rare-earth metal were carried out there (Bjerrum Møller *et al* 1966, 1968).

When the reactor at the Institut Laue–Langevin began operating in 1972, a revolution in neutron scattering was brought about. Whole new areas of (Q, ω) space were opened up to entirely new communities of scientists.

A possible entry for the later years of this period is discussed in § 9.3 by R M Moon in his account of the Center for Neutron Research that is being designed as a new resource in the United States.

References

Alverez L and Bloch F 1940 *Phys. Rev.* **57** 111
Bacon G E 1951 *Proc. R. Soc.* A **209** 397–407
Bloch F 1936 *Phys. Rev.* **50** 259
—— 1937 *Phys. Rev.* **51** 994
Bjerrum Møller H and Gylden Houmann J C 1966 *Phys. Rev. Lett.* **16** 737
—— 1968 *J. Appl. Phys.* **39** 807
Brockhouse B N 1957 *Phys. Rev.* **106** 859
—— 1958 *Phys. Rev.* **111** 1273
Collins M F, Minkiewicz V J, Nathans R, Passell L and Shirane G 1969 *Phys. Rev.* **179** 417
Elliott R J and Lowde R D 1955 *Proc. R. Soc.* A **230** 46
Gersch H A, Shull C G and Wilkinson M K 1956 *Phys. Rev.* **103** 516
von Halban H, Frisch O and Koch J 1937 *Nature* **139** 756, 1021
—— 1938 *Phys. Rev.* **53** 719
Halpern O and Johnson M H 1939 *Phys. Rev.* **55** 898–923
Hoffman J G, Livingston M S and Bethe H A 1937 *Phys. Rev.* **51** 214
Hughes D J and Burgy M T 1951 *Phys. Rev.* **81** 489–506
Jacrot B, Konstantinovic J, Parrette G and Cribier D 1963 *Inelastic Scattering of Neutrons in Solids and Liquids* vol II (Vienna: IAEA) p 317–26 and references therein
Latham R and Cassels J M 1952 *Proc. Phys. Soc.* A **65** 241
Lovesey S 1984 *Theory of Neutron Scattering from Condensed Matter* (Oxford: Clarendon) Chap 13

Lowde R D 1954 *Proc. R. Soc.* A **221** 206–23

Marshall W and Lovesey S W 1971 *Theory of Thermal Neutron Scattering* (Oxford: Clarendon) Chap 3 and Appendix B

Marshall W and Lowde R D 1968 *Rep. Prog. Phys.* **31** 705

Migdal A 1938 *C. R. Acad. Sci. l'URSS* **20** 551 (this article was brought to my attention by Dr V F Sears)

Mook H A and Lynn J W 1984 *Proc 30th Annual Conf on Magnetism and Magnetic Materials San Diego, Ca.* in press

Moon R M, Riste T and Koehler W C 1969 *Phys. Rev.* **181** 920–31

Néel L 1936 *C. R. Acad. Sci., Paris* **203** 304

Palevsky H and Hughes D J 1953 *Phys. Rev.* **92** 202–3

Passell L, Blinowski K, Brun T and Nielsen P 1964 *J. Appl. Phys.* **35** 933–4

Peterson S W and Levy H A 1951 *J. Chem. Phys.* **19** 1416–8

Powers P N *et al* 1937 *Phys. Rev.* **51** 51, 371, 1112

—— 1938 *Phys. Rev.* **54** 827

Riste T, Blinowski K and Janik J 1959 *J. Phys. Chem. Solids* **9** 153; see also Goedkoop J A and Riste T 1960 *Nature* **185** 450

Schofield P (ed) 1982 *The Neutron and its Applications* (Inst. Phys. Conf. Ser. 64)

Schwinger J 1937 *Phys. Rev.* **51** 544

Shirane G 1984 *J. Magn. Magn. Materials* **45** 33

Shull C G 1951 *Phys. Rev.* **82** 626

Shull C G and Smart J S 1949 *Phys. Rev.* **76** 1256; see also Shull C G, Strauser W A and Wollan E O 1957 *Phys. Rev.* **83** 333–45

Shull C G, Wollan E O and Koehler W C 1951 *Phys. Rev.* **84** 912–21

Sinclair R N and Brockhouse B N 1960 *Phys. Rev.* **120** 1638–40

Squires G L 1954 *Proc. Phys. Soc.* A **67** 248–53

Steinberger J and Wick G C 1949 *Phys. Rev.* **76** 994–5

Tucciarone A, Lau H Y, Corliss L M, Delapalme A and Hastings J M 1971 *Phys. Rev.* B **4** 3206

Van Hove L 1954a *Phys. Rev.* **95** 249–62

—— 1954b *Phys. Rev.* **95** 1374–84

Wilkinson M K and Shull C G 1956 *Phys. Rev.* **103** 516–24

Wollan E O and Shull C G 1948a *Phys. Rev.* **73** 830–41

—— 1948b *Nucleonics* **3** 17

7.2 Neutrons and Chemistry

H Fuess

Institut für Kristallographie und Mineralogie der Universität, Frankfurt am Main, Federal Republic of Germany

Introduction

Research in chemistry is mainly concerned with the preparation of new molecules and compounds. The characterisation of material and the investigation of relevant properties by spectroscopic and diffraction methods is an additional very important task of chemical research. Both diffraction and spectroscopy may be performed by neutron scattering due to the simultaneous occurrence of energy and momentum transfer during the process of scattering neutrons. Whereas diffraction reveals the static arrangement of atoms in crystals, spectroscopy is concerned with dynamic processes like the motion of molecules in condensed matter. The instrumentation for diffraction was developed along the successful lines known from X-ray diffraction. The facilities to measure energy transfer in different energy ranges were in most cases specially adapted to a particular application and built for the purpose.

The application of neutrons to chemical research started with diffraction experiments in solid-state chemistry at room temperature. The experiments now extend to liquids, glasses, polymers, solutions and surface studies and the sample environment now covers temperatures from some mK to about 2300 °C, pressures up to several kilobars and magnetic fields of some tesla.

The purpose of this contribution is to give an account of the role of neutron scattering in all areas of chemistry. We shall try to outline briefly the historical development in several fields and describe the new knowledge which has been obtained. We shall start with results from elastic scattering in condensed matter and then add knowledge from inelastic experiments.

Structure Determination by Neutron Diffraction

Inorganic material

Already the first neutron diffraction experiment clarified a controversial point in solid state chemisty. Shull *et al* (1948) determined the structure of sodium hydride and deuteride unambiguously from powders and found that both compounds crystallise in the NaCl-type structure.

Research in the decade 1950–60 saw the structure determination of many hydrogeneous compounds and thus provided a solid base for a discussion of the hydrogen bond. Ice, crystalline hydrates and acid salts were among the first substances successfully studied. A powder study at Oak Ridge on hexagonal ice already indicated the correct split position of hydrogen, nicely demonstrated in the single-crystal work of Peterson and Levy (1957). Recently a detailed programme on the polymorphism of ice was launched at the Institut Laue–Langevin, Grenoble. This investigation is based on powder as well as single-crystal samples and covers the temperature range down to 4 K and pressure up to several kilobars to determine clearly the complicated phase diagram of ice.

This continuous interest in ice illustrates the progress in techniques and methods from 1948 to the present day and demonstrates some of the main advantages of neutrons in crystal chemistry: (i) determination of hydrogen positions, (ii) relatively easy use of ancillary equipment like cryostats and pressure cells and (iii) the strong role played by powder diffraction, essentially after its revival by the profile analysis technique introduced by Rietveld (1969) which placed neutron powder diffraction well ahead of similar X-ray techniques. Despite this relative strength of powder diffraction very impressive results were obtained by single-crystal diffractometry which used in most cases 4-circle diffractometers.

A recent result of that technique is shown in figure 7.1. A single hydrogen atom is clearly detected in the middle of a cage of ruthenium atoms (Jackson *et al* 1980). This is a good demonstration of the power of neutrons to determine structural details not available to X-ray diffraction.

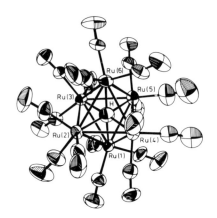

Figure 7.1 The structure of the $HRu_6(CO)_{18}^-$ anion complex (Jackson *et al* 1980).

Structure determination of organic molecules

When we consider organic compounds we are inevitably more concerned with discrete molecules. The contribution of neutrons was not only the precise determination of hydrogen positions but also the evaluation of molecular motion. Motion was studied in some cases by elastic (Debye–Waller factor) and inelastic scattering on the same compound. One of the first organic compounds studied with neutrons was oxalic acid dihydrate which seems to constitute the subject of another never-ending story in diffraction history. The same compound was among the first molecules investigated for deformation densities and quite recently a project of the International Union of Crystallography was focused on oxalic acid which therefore seems to be the compound on which most neutron diffraction data sets were collected. The hydrogen bonding scheme of α-resorcinol $(m-C_6H_4(IH)_2)$ resulted from the first study of an aromatic compound which was performed at Harwell in the United Kingdom by Bacon and Curry (1956).

Systematic crystal structural studies were performed on several classes of organic compounds. They include aliphatic and aromatic hydrocarbons, amino acids and carbohydrates. It was demonstrated that the simplest organic molecule, methane, has a very complicated series of phase transitions in the solid state including plastic phases with rotating CH_4 molecules. The degrees of freedom and the motion were determined simultaneously by inelastic scattering (see Press 1981).

The study of amino acids by the late Walter Hamilton at Brookhaven in the 1960s provided a wealth of precise structure determinations and a solid base for discussion of the hydrogen bonding schemes in these constituents of proteins. This work on amino acids was summarised and discussed in some detail by Koetzle and Lehmann (1976). Additional information on amino acids was obtained at Trombay (India).

The study of carbohydrates (pyranoses and disaccharides) at Brookhaven from 1974 onwards led Jeffrey (1978) and his co-workers to the formulation of general rules for hydrogen bonding in these types of compound. These results stimulated theoretical work which led to a reinterpretation of the nature of the hydrogen bonds in these systems.

The crystal structure investigation of vitamin B12 at Harwell by Hodgkin, Willis and others in the 1960s was an outstanding achievement and a pioneering work due to the relative large size of the molecule and the rather limited flux of the DIDO reactor. Nevertheless an understanding of the water–molecule interaction was obtained.

Neutron powder diffraction

As already mentioned, most of the early neutron diffraction work was performed on powder samples. We think that it is therefore worthwhile to outline the development in that area. A large and important field of investigation is formed by hydrogen in metals. This is a good example where a combination of elastic and inelastic scattering provided a base for a deeper understanding of the diffusion processes of hydrogen in different metals.

Although the location of hydrogen atoms has proved one of the most attractive applications for neutron diffraction the contribution to heavy element chemistry by locating carbon, nitrogen and oxygen must not be overlooked. Among the first oxides were La_2O_3, Sc_2O_3, U_3O_8 and UO_2, PbO and AgO. The structure investigation of UO_2 and the incorporation of additional oxygen in this fluorite-type structure was among the first non-stoichiometric compounds which were characterised by neutrons. Further studies on cation deficient oxides like $Fe_{1-x}O$ and solid solutions of YF_3 in CaF_2 led to the formulation of 'clusters' of interstitial atoms and defects in non-stoichiometric compounds. The determination of vacancies in carbides should be mentioned in this context, too.

Magnetic structures. Magnetic structure determination constituted one of the earliest domains of successful application of neutrons. Magnetism may be considered as a border region between chemistry and physics; nevertheless the knowledge of magnetic structures, and their classification into magnetic space groups, is of great importance to crystallography and magnetochemistry and even contributes towards an understanding of the chemical bond. The studies of the transition metal oxides MnO, FeO, CoO and NiO were among the first neutron powder patterns. This work performed at Oak Ridge led to the determination of the antiferromagnetic spin arrangement in these compounds. This was followed by the study of ferrimagnetic spinels and perovskites at Oak Ridge and Brookhaven (Corliss and Hastings) and at the French nuclear research reactors at Saclay (Mériel) and Grenoble (Bertaut) which became operational about 1960. The enormous amount of information produced by these groups (especially in the decade 1960–70) induced the complete formulation of magnetic space groups and the concept of magnetism and the chemical bond by Goodenough (1963), who formulated rules for the concept of superexchange interactions in insulators.

Powder diffraction was successfully applied as magnetic reflections occur only at low angles where the resolution of diffractometers was sufficient to establish the magnetic structure. The powder method was, however, limited for general application because of the overlap of reflections and the low resolution of powder diffractometers.

Progress in powder diffraction. The shortcoming of this technique was overcome when Rietveld introduced the profile analysis technique which made full use of the entire pattern. Each point is regarded as an individual observation and the crystallographic parameters are refined to reproduce the pattern as a whole. This technique opened new areas to neutron powder diffraction. In particular, structures of low symmetry and structural phase transitions were now investigated. The progress in the analysis technique called for diffractometers with higher angular resolution.

On steady-state reactors the resolution was obtained by high take-off angles and improved collimation. Good examples are the diffractometers D1A and D2B at the ILL, Grenoble (Hewat 1976). The advent of pulsed neutron sources allows full use of back-scattering in connection with time-of-flight techniques. Due to these improvements the accessible range in neutron powder patterns is considerably extended. High resolution powder spectra down to d-values of 0.3 Å were successfully refined.

Due to higher resolution instruments powder diffraction on larger organic molecules became feasible. Powder diffraction work in this field started with small hydrocarbons like methane or acetylene with special emphasis on the study of phase transitions.

Profile analysis together with constrained refinement techniques allowed the refinement of complete aromatic structures. The molecule is treated as a rigid unit and its position in a unit cell is determined by only very few parameters which are the mean bond lengths and the angles between the rigid unit and the crystollographic axes. Structures refined include bromine and iodine derivatives of benzene. Even complete structure determinations were achieved without previous knowledge of parameters from X-ray work. On polyethylene a refinement of structural parameters from powder data was reported and demonstrates the high standard.

Chemical reaction in the solid state. The installation of two-dimensional detectors and fast data-retrieval systems made the direct observation of chemical reactions in solid state chemistry possible. One of the first observations of this type was the formation of intercalation compounds of TaS_2 and NH_3. Riekel (1980) described the appearance and disappearance of powder lines as a function of time and gave an indication of the various stages of formation which occur during the reaction.

A very recent example is illustrated in figure 7.2 where the formation of concrete from $CaO(C)$, $SiO_2(S)$ and D_2O is shown in a real-time experiment (Christensen *et al* 1985). A detailed research programme studying reactions at different temperatures and with a range of composition may elucidate the complicated phase diagram of concrete.

Hydrogen bonding
One leitmotiv of neutron diffraction studies was hydrogen bonding, with first investigations of the short hydrogen bonds in KHF_2 and KH_2PO_4. The determination of hydrogen positions in some sulphates ($CuSO_4.5H_2O$; $CaSO_4.2H_2O$, $Li_2SO_4.H_2O$, potassium chromium alum) and several other hydrates up to 1960 gave evidence on the deformation of the water

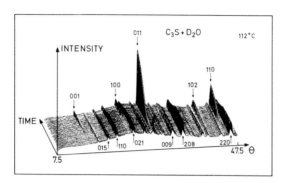

Figure 7.2 Crystallisation of concrete. Powder diffraction pattern as a function of time (Christensen *et al* 1985).

molecule in the crystalline state. The main result is a larger H–O–H angle than observed in unbonded free water and slightly longer O–H distances. The experimental results on hydrates were summarised by Ferraris and co-workers (1982).

The wealth of information on hydrogen bonds where oxygen acts as donor and acceptor is summarised in figure 7.3 which represents the correlation between O–H and O...O distances. This figure has two distinct branches, one for symmetrically restricted short hydrogen bonds, the other one for all H bonds. The plot suggests that symmetrical and asymmetrical bonds may coexist for the same O...O distance: most hydrogens in these short bonds can be interpreted either in terms of a centred model with anisotropic thermal motion or by a disordered model with two half hydrogens. The question of true symmetrical H bonds as discussed by Bacon in 1963 is therefore still open. Additional evidence on that question may come from inelastic studies.

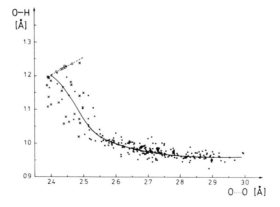

Figure 7.3 Correlation between O–H and O...O distances in O–H...O hydrogen bonds.

A description of hydrogen positions in solids can not be concluded without mentioning the investigation of the phase changes in ammonium halides by Levy and Peterson (1953) who demonstrated in this classical study the importance of hydrogen bonding for the various phases of NH_4Br, NH_4Cl and NH_4I (figure 7.4).

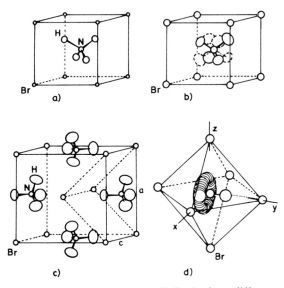

Figure 7.4 The crystal structure of ND_4Br in four different phases (Levy and Peterson 1953).

Electron and Magnetisation Densities

The distribution of electrons between atoms is considered to be a sign of chemical bonds. This density is of course studied by X-ray diffraction. Considerable progress was made in the field when Coppens introduced a combination of X-ray and neutron diffraction, a difference method called the X–N technique. The total electron density is determined by X-ray diffraction, the precise position of the nucleus and its thermal motion by neutron diffraction. A calculation of an electron density at the atomic position with spherical form factors for an unbounded atom produces a promolecule. The difference map is obtained by subtracting the promolecule density from the total density. The X–N studies which were published in the last decade on a number of organic molecules (oxalic acid, urea, thriazine, tetracyanoethylene and others) provided evidence of accumulation of electrons in C–H, C–O and C–C bonds. This evidence was supported by *ab initio* molecular orbital calculations. Figure 7.5 gives an

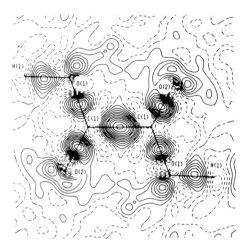

Figure 7.5 Deformation electron density (X–N technique) of oxalic acid dihydrate (IUCr 1984).

experimental deformation density of the oxalic acid molecule as a result of a common effort of many authors (IUCr 1984). Astonishing results were obtained for metal–metal bonds and for H_2O_2, where no density was found between atoms. These results made some investigations necessary to reconsider the very nature of the chemical bond.

A direct determination of the distribution of 3d and 4f electrons in magnetically ordered compounds is achieved by the precise measurement of magnetic structure factors. The technique of polarised neutrons is applied for this purpose. The first experiments on Fe, Co and Ni at Brookhaven and Oak Ridge after 1960 demonstrated the delocalisation of d electrons in these metals. The study of Wedgwood (1976) on the $(CrF_6)^{3-}$ ion in NaK_2CrF_6 revealed directly, for the first time, unpaired spins on a non-magnetic ion (fluorine) and therefore gave clear evidence of the covalent contribution to that bond. This contribution was observed on maps of spin densities obtained by Fourier inversion of the magnetic structure factors. Magnetic densities in yttrium iron garnet $Y_3Fe_3O_{12}$ and in Fe_3O_4 indicated uncancelled spin on oxygen and thus illustrated the superexchange between the magnetic ions via the ligand.

A more quantitative interpretation of polarised neutron studies was introduced by Chandler *et al* (1982). They analysed the data on several complex ions composed of a 3d element and halides by molecular orbital theories. Thus an occupation number could be derived for each d orbital. In $Ni(NH_3)_4(NO_2)_2$ the magnetisation and the electron density were analysed simultaneously and the occupation of d orbitals was obtained by both techniques.

Polymer Studies

Neutron scattering has contributed considerably to the understanding of polymers. Whereas small- and low-angle scattering gave evidence on the structure of polymer molecules in solution and in the solid state, the study of energy transfer was instrumental in the interpretation of dynamical processes in polymers. As most polymers contain huge numbers of hydrogen atoms a great gain arises because of the very different scattering lengths of hydrogen and deuterium. A single molecule may thus be 'coloured red' by complete deuteration. The contrast in scattering then gives information on the conformation of this particular molecule. First neutron scattering experiments were carried out at the FRJ2 reactor at Jülich by Kirste *et al* (1972) and led subsequently to exciting work on concentrated solutions, networks and gels, on crystal topology and orientation. The use of D-containing polymers in their H-containing counterparts was outstandingly successful. Neutron diffraction gave experimental evidence of a random-walk structure instead of a regular folded structure for polymers in the melt. The details obtained from solution were the molecular weight and the conformation of single chains. Furthermore the radius of gyration indicates the approximate dimension of a single molecule. In dilute solutions the polymers behave rather like spheres.

The isomorphous replacement method was also valuable for structure investigations in the solid state. For polyethylene the so-called solidification model was derived (Stamm *et al* 1979). It assumes that the structure in the solid occurs by straightening coil sequences which are already present in the melt without long range diffusion processes (figure 7.6). Here a clarification of a controversial subject was provided by neutron scattering.

Early concepts of polymer crystals like the so-called 'fringed micelle' models in which chains were supposed to pass through crystalline regions and back into the amorphous phase could be definitely ruled out. Additional experiments on stretched polymers were carried out which brought some conclusions on their elasticity. Conducting polymers (like

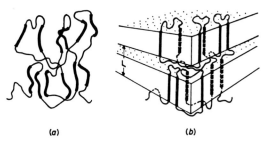

(a)   (b)

Figure 7.6 Structure of polyethylene (*a*) in the melt (*b*) in the solid state (Stamm *et al* 1979).

polyparaphenyl doped with AsF_5) and their phase changes were studied quite recently.

Extensive studies by inelastic neutron scattering on the diffusion of polymers in solution are just at their beginning on polyethylene, polystyrole, polypropylene and others. Different models are developed and tested and the outcome will definitely contribute to the understanding of these synthetic macromolecules. Inelastic scattering studies already explain diffusive motion within one chain. It is, however, not yet possible to give a quantitative interpretation of the melt on a molecular level.

Liquids, Solutions and Amorphous Material

The first neutron diffraction studies on liquids were performed in 1950 by Chamberlain who examined liquid samples of sulphur, lead and bismuth Convincing results on the ordering in liquids were obtained on melts of metals and alloys and on the inert gases Ar, Ne, Kr and He which produced a new understanding of forces acting in the liquid and gaseous state.

The essential quantity for a quantitative description of the liquid state is the space–time correlation function. It defines the probability of finding an atom at a given place. The interpretation of first results on liquid metals (Zn, Pb, Tl, Sb and Bi) in 1968 showed that a hard sphere model is not sufficient for the interpretation of the data obtained. More sophisticated potentials had to be used instead.

When the study of simple monatomic liquids is extended to species with different atoms, like simple alloys, it is found that neutrons have a very significant advantage because isotopic replacement allows different atoms to be distinguished. Therefore many more pair-correlation functions may be derived.

This method has been extremely successful in the investigation of solutions which are rather complicated systems. An aqueous solution of a diatomic salt (e.g. NaCl or $CaCl_2$) contains four different atoms and therefore already ten different interactions have to be considered. This difficulty may explain why the first systematic studies were not reported prior to 1970 when Enderby and co-workers (Cunnings *et al* 1980) introduced the method of isotopic replacement. The diffraction pattern with ^{37}Cl and ^{35}Cl and the use of different isotopes of Ca and Na in both H_2O and D_2O gave sufficient information to deduce all ten pair-correlation functions and to construct a model for the hydration spheres around the ions in solution. These investigations provided for the first time evidence for a second coordination sphere around the cation. Futhermore they demonstrated that different cations have little or no influence on the hydration sphere around Cl^- which is quite similar for all the chlorides studied. A good deal of information on the structures of liquids was obtained from neutrons because X-ray studies often relate to heavily contaminated surfaces.

Surface Investigations

The study of surface effects by neutron scattering is one of the latest examples of successful application of neutrons to chemical problems. The results may be subdivided into the static arrangement of absorbed molecules on the surface, on the one hand, and in the determination of vibrational energies on the other. The first studies on surface structures were performed at about 1972 and this field is still of growing interest (White 1977). The systems investigated consisted of monolayers of gas molecules (e.g. N_2, Ar) on metal surfaces or graphite.

The structure of the monolayer does in some cases coincide with the hexagonal arrangement of graphite, sometimes the periodicity is different. A quite widespread arrangement is the $\sqrt{3} \times \sqrt{3}$ structure which has a unit cell of the monolayer which is $\sqrt{3}$ times bigger than the graphite cell. In addition to structure determinations the energies of absorption and vibration of absorbed molecules were investigated by inelastic neutron scattering. The vibrational spectrum of benzene on finely dispersed nickel (Raney nickel) is very different from that of pure benzene and reflects the energies of physisorption. Important contributions to the understanding of catalytic processes are to be expected from similar studies of vibrational energies of hydrocarbons on the surfaces of platinum, nickel or graphite or of the diffusion of small molecules in the voids of zeolitic material.

Molecular Vibration, Diffusion and Rotation

The field of molecular motion may be subdivided into (i) large energy changes in the neutron spectrum which are due to fast motions of molecules or molecular fragments and (ii) small changes as a result of slow diffusion processes.

Early examples of the relevance of inelastic neutron scattering to molecular motion in crystals were given by the spectroscopy of ammonium salts. The study of $(NH_4)_2SO_4$ was intended to detect energy changes at the ferroelectric phase transition. The absence of a drastic energy change is evidence for a structural phase transition rather than an onset of free rotation. Since the advent of high-resolution instruments, especially the backscattering technique after 1973, energy resolution below 1 meV became feasible. Thus single rotation in molecular crystals became a broad field of application.

Freely rotating NH_4^+ groups and different forms of hindered rotations were extensively studied in ammonium halides and proved the correctness of the early structural models of Levy and Peterson (1953). Inelastic spectra on NH_4Br are shown in figure 7.7. They show the evolution of the quasielastic peak at the bottom of the elastic line as a function of temperature (Lechner *et al* 1980). Besides the NH_4^+ group other molecules, like methane (CH_4 or CD_4), nitrogen (N_2) or hydrogen (H_2) in their solid state and molecular fragments like $-CH_3$ and $-NH_3$ groups were investigated. It was demonstrated that a continuous transition exists from quantum mechanical rotation at low temperatures (characterised by definite tunnel frequencies) to classical diffusive rotational motion at higher temperatures. In neutron scattering this means a transition from line spectra at low temperature to quasielastic scattering at high temperature.

An interesting intermediate case is a rotational jump model which assumes that the jump time required for a diffusive step is much shorter than the residence time at a given position. This model adequately describes the motion of $(NH_4)_2SnCl_6$ or $Pb(CH_3)_4$, as well as the ammonium halides in some phases or the motion of $-NH_3$ in anilinium bromide. This jump rotational model is reflected in a broadening of the quasielastic line.

The temperature dependence of the intensity of the quasielastic line gives the activation energy. For several compounds the jump model becomes inadequate (residence time longer than jump time) and it should be replaced by a model accounting for continuous diffusion in a given potential.

The interpretation of the quasielastic line is, however, difficult because of the need to separate elastic and quasielastic spectra. An example for a jump diffusion model is given by the mobility of Ag^+ ions in the solid electrolyte AgI. Similarities of the motion of Ag^+ to that of H in metals are obvious. In both cases a model with residence times longer than the time for a jump from one position to the next describes the quasielastic line fairly well.

The motion of molecules (methane, adamantane etc, and ions like NH_4^+) in solids is of continuing interest in inelastic neutron scattering and

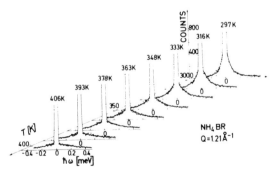

Figure 7.7 Temperature dependence of the quasielastic scattering in NH_4Br (Lechner *et al* 1980).

constitutes another never-ending story, where early work was repeated with higher energy resolution and with more sophisticated interpretations. Quite often early powder work was superseded by single-crystal work which allowed relation of the molecular motion to crystallographic axes. Furthermore deuteration or partial deuteration of groups of molecules (e.g. a CH_3 group) helped considerably in attributing energy lines.

Conclusion

Neutron scattering research in topics related to chemistry covers a wide range from structure investigation in crystalline solids, through glasses, surfaces and solutions to molecular spectroscopy and diffusion processes. A few examples have been chosen to trace the development from the first neutron diffraction experiment to current fields of research. A detailed review of results from neutron diffraction is given by the present author (Fuess 1979), and more detailed descriptions of inelastic scattering results appear in Press (1981) and White (1977). The whole field is still in active development and many more results are expected.

References

Bacon G E 1963 *Applications of Neutron Diffraction in Chemistry* (Oxford: Pergamon)

Bacon G E and Curry N A 1956 *Proc. R. Soc.* A **235** 552

Chamberlain O 1950 *Phys. Rev.* **77** 305

Chandler G S, Figgis B N, Phillips R A, Reynolds P A and Mason R 1982 *J. Physique Suppl.* C **7** 323

Chiari G and Ferraris G 1982 *Acta Crystallogr.* B **38** 2331; see also Joswig W, Fuess H and Ferraris G 1982 *Acta Crystallogr.* B **38** 2798

Christensen A N, Fjellvag H and Lehmann M S 1985 *Acta Chem. Scand.* a **39** 593

Cunnings S, Enderby J E and Howe R A 1980 *J. Phys. C: Solid State Phys.* **13** 1

Fuess H 1979 *Modern Physics in Chemistry* ed E Fluck and V I Goldanski (London: Academic) pp 1–193

Goodenough J R 1963 *Magnetism and the Chemical Bond* (New York: Wiley)

Hewat A W 1976 *Nucl. Instrum. Methods* **137** 463

International Union of Crystallography 1984 *Acta Crystallogr.* A **40** 184

Jackson P F, Johnson B F G, Lewis J, Raithby P R, Mapartlin M, Nelson W J H, Rouse K D, Allibon J and Mason S A 1980 *J. Chem. Soc. Chem. Commun.* 295

Jeffrey G A and Takagi S 1978 *Acc. Chem. Res.* **11** 264 (see also Jeffrey G A, McMullan R K and Takagi S 1977 *Acta Crystallogr.* B **33** 728)

Kirste R G, Kruse W A and Schelten J 1972 *Makromolek. Chem.* **162** 299

Koetzle T F and Lehmann M S 1976 *The Hydrogen Bond* ed P Schuster, G Zundel and E Sandorfy (Amsterdam: North-Holland) pp 457–65

Lechner R E, Badurek G, Dianoux A J, Hervet H and Volino F 1980 *J. Chem. Phys.* **73** 934

Levy H A and Peterson S W 1953 *J. Chem. Phys.* **21** 366
Peterson S W and Levy H A 1957 *Acta Crystallogr.* **10** 70
Press W 1981 *Single Particle Rotations in Molecular Crystals* (Heidelberg: Springer)
Riekel C 1980 *Prog. Solid State Chem.* **13** 89
Rietveld H M 1969 *J Appl. Crystallogr.* **2** 65
Shull C G, Wollan E O, Morton G A and Davidson W C 1948 *Phys. Rev.* **73** 842
Stamm M, Fischer E W, Dettenmaier M and Convert P 1979 *Faraday Disc., R. Soc. Chem.* **68** 263
Wedgwood F A 1976 *Proc. R. Soc.* A **349** 447
White J W 1977 *Dynamics of Liquids by Neutron Scattering* (Heidelberg: Springer)

7.3 Neutrons and Lattice Dynamics

B M Powell and A D B Woods

Chalk River Nuclear Laboratories, Ontario, Canada

The Beginning

In the beginning was an Idea—that from measurements of the energy and momentum changes of thermal neutrons scattered by crystalline solids, the lattice dynamical dispersion relation could be directly determined. This idea appears to have occurred to several people independently and more or less simultaneously in the early 1950s. The concept of lattice dynamics, the collective vibrations of atoms or molecules in a solid, was developed many years before the discovery of the neutron and many properties can be analysed in terms of these collective excitations or 'phonons'. It was realised that lattice dynamics could be used as a probe of the fundamental interatomic forces, but this would require the unique energy and momentum relation of the phonons to be measured, i.e. the phonon dispersion surface mapped out. The coherent, inelastic scattering of thermal neutrons promised to be a powerful technique for such measurements and the challenge was to turn this possibility into reality.

The properties which give thermal neutrons this unique ability are the almost perfect match between the wavelength and energy of the neutron and the corresponding characteristics of typical lattice vibrations. Consequently, for an event in which a thermal neutron is scattered by a phonon, the former suffers large (and so easily measurable) changes in its energy and momentum. These changes measure *directly* the energy and momentum of the phonon. Thus by making a sequence of neutron scatter-ing measurements, the entire phonon dispersion surface may be mapped out. A further property of thermal neutrons is that they are a weak probe and so explore the *bulk* properties of the solid without themselves perturb-ing the system.

Despite these favourable properties, there were several other require-ments to be met before the idea could become a reality. A beam of incident neutrons well defined in both energy and momentum, a method to mea-sure the energy and momentum of the scattered neutrons and a mechanism to orient single-crystal specimens were all necessary. Last, but not least, a high-intensity source of thermal neutrons was required. The early 1950s saw the development of several neutron spectrometers designed to achieve some or all of the above requirements. By the mid-1950s two basic types of spectrometer had evolved; time-of-flight spectrometers in which the neut-ron energy was measured by timing the neutron flight over a known distance, and crystal spectrometers in which the energy was measured by Bragg-reflecting the neutron through a known angle from a single crystal. The year 1955 saw phonon results from Chalk River, Saclay and Brookhaven. At Chalk River, Brockhouse and Stewart made the first measurement of a phonon dispersion curve, that for [111] transverse modes in Al as shown in figure 7.8, and so unequivocally demonstrated the existence of well defined phonons in a metal. In 1958 the same authors published the dispersion curves of Al for all three high-symmetry direc-tions, the most complete set of data measured to that time. These data were sufficiently detailed to allow meaningful discussion of the most

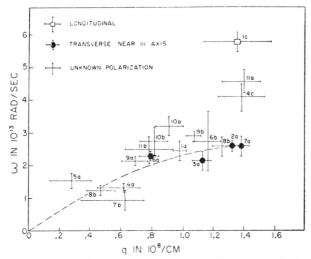

Figure 7.8 The first phonon dispersion curves. They were obtained for aluminium at Chalk River (Brockhouse B N and Stewart A T 1955 *Phys. Rev.* **100** 756).

appropriate description of the interatomic forces in Al. In the same year Brockhouse and Iyengar measured dispersion curves for two of the symmetry directions in germanium which, since there are two atoms in the primitive unit cell, has both acoustic and optic branches in its dispersion relation. They successfully applied the concept of the 'one-phonon dynamical structure factor' to assign their data to acoustic or optical branches, the first recognition of the need for structure factor calculations for all but the simplest of crystals. The following year Woods, Cochran and Brockhouse extended the technique to crystals with two different atoms in the unit cell when they reported dispersion curve measurements for NaI. The data were analysed in terms of a lattice dynamical model which successfully included atomic polarisability—the shell model.

The End of the Beginning

The year 1960 was a landmark year in the field for it saw the first of the series of IAEA sponsored meetings on inelastic neutron scattering. New results were reported from several laboratories, but more important it was at this meeting that Brockhouse described in detail the constant-Q mode of operation of the triple-axis spectrometer (see figure 7.9). It was already evident that analysis of the phonon dispersion curves was significantly easier if the measurements were made along directions of high symmetry

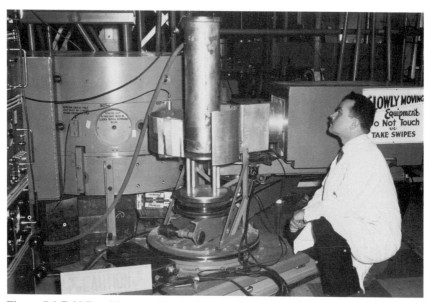

Figure 7.9 B N Brockhouse with the first version of his triple-axis spectrometer at the NRU reactor, Chalk River, in 1959.

and that particular momentum transfers were often of special interest. It was thus clear that it would be advantageous if the configuration of the neutron spectrometer could be chosen so as to make measurements at specified momentum transfers and the ability to choose specific configurations was much more easily achieved with a crystal spectrometer than with a time-of-flight spectrometer. It was this capability which led to the success of the triple-axis spectrometer and the invention of the constant-Q mode of operation which allowed a specified range of frequency to be investigated at selected momentum transfers. This invention was the most important instrumental development in the application of inelastic neutron scattering to lattice dynamics and the triple-axis spectrometer quickly became the dominant instrument in the field.

The number of laboratories carrying out inelastic neutron-scattering measurements and the range of materials under investigation increased steadily in the early 1960s. The face-centred cubic elemental metals were among the first materials to have their dispersion curves measured and the data were usually analysed in terms of the Born–von Kármán formalism. The case of lead was of special interest, for not only was it the first material whose dispersion curves were measured by the constant-Q technique, but it was the first material in which the effects of temperature dependent anharmonicity and the electron–phonon interaction were seen in the dispersion curves. Of the body-centred cubic metallic elements the alkali metals were of special interest because their relatively simple electronic structures encouraged theoretical calculations for comparison with the neutron results. Several of the transition metals proved to have remarkably complex dispersion curves with large anomalies attributed to the electron–phonon interaction. These data helped to stimulate efforts to understand, from first principles, electron–phonon effects in these complex metals. Phonons in the hexagonal close-packed elements were also intensively studied while measurements on the elements with lower symmetry, e.g. Bi, Te and Sn were made rather later.

The Middle

Another major watershed was reached in 1965 when two new high-flux reactors came on line—HFIR at Oak Ridge and HFBR at Brookhaven. The majority of instruments at both these reactors were triple-axis spectrometers and, combined with the higher thermal flux of these sources, provided a huge increase in the available measuring power for collective excitations in general and lattice dynamics in particular. In the same year the first overall review of the practice of neutron scattering appeared (Thermal Neutron Scattering, edited by P A Egelstaff (1965)) and the article by Dolling and Woods on 'Thermal Vibrations of Crystal Lattices' reviewed its application to the measurements of lattice dynamics for the materials studied to that date. In 1972 a further significant increase in the availability of thermal neutrons occurred when the high-flux reactor at Institut Laue–Langevin, Grenoble began operation. Although many of the instruments at this institute were designed for highly specialised measurements, several triple-axis spectrometers were also constructed. In 1974 another review of the practice of neutron scattering applied to lattice dynamics was given by Dolling in the chapter 'Neutron Spectroscopy and Lattice Dynamics' in Volume 1 of *Dynamical Properties of Solids,* edited by Horton and Maradudin (1974). The review contained a comprehensive list of materials for which neutron data existed to that date, perhaps the last time such a compilation will be made in a review format. For, as the use of neutron scattering has become more widespread, the range of materials studied has grown enormously and the results are increasingly directed to specialists in the particular field under investigation, for example a useful encyclopedia *Phonon Dispersion Relations in Insulators* was compiled in 1979 by Bilz and Kress.

During this period many measurements of dispersion curves of ionic and covalently bonded materials such as the alkali halides, elemental semiconductors, diatomic semiconductors and insulators and many other diatomic compounds, were made and often interpreted in terms of the shell model. The concepts of this model were used by Cochran to develop a theory of ferroelectric phase transitions which became known as the 'soft-mode' theory. In this theory the ferroelectric transition occurs when the frequency of a temperature-dependent transverse optic mode approaches zero. The crystal structure then becomes unstable with respect to the displacement pattern of this mode and makes a transition to a more stable structure. The first observations of such a mode were made in the perovskite $SrTiO_3$ by Cowley at Chalk River in 1962. The success of this soft-mode theory led to extensive studies of other ferroelectrics and phase transitions in general, particularly at Brookhaven in the late 1960s. The concept of a phase transition arising as the result of the displacement pattern of some mode becoming 'frozen-in' at the transition temperature has since been widely used in interpreting structural phase transitions. Although under detailed investigation many transitions are found not to be ideal soft-mode transitions, the model has proved invaluable in interpreting transitions ranging from soft rotary modes in antifluorites to incipient ω-phase formation in transition metals.

Neutron scattering measurements of 'defect' systems, in which a low concentration of guest atoms or molecules are incorporated into a host lattice, were also popular but generally proved difficult. Depending on the mismatch between the masses and force-fields of the host and guest one may observe perturbations of the host phonons, additional modes or coupled modes of the guest and host. Most of the results were obtained for metallic systems, e.g. Cu(Au), Cr(W), but some non-metallic systems were investigated e.g. SnTe–GeTe. There has been rather more success in lattice dynamical measurements of high-concentration mixtures and of alloy systems, for the dispersion curves often show significant shifts from those of the pure materials but remain well defined. However, the lineshapes usually contain structure which is difficult to measure accurately with the experimental resolution currently available.

Among the simplest of crystals are the solid rare-gases and so these were obvious candidates for comparison of lattice dynamical measurements with theoretical calculations. Once single crystals became available, extensive measurements of the phonon dispersion curves in the heavy rare gases were made, principally at Brookhaven. The temperature and pressure dependences of the phonons were measured and effects attributable to three-body interactions also explored. These results were followed by measurements on single crystals of the quantum solid ^{4}He. Dispersion curves were measured in all three phases of solid ^{4}He at Brookhaven and Ames and the data proved invaluable in testing theories of the dynamics of quantum solids. A review of the neutron scattering results has been given by Powell and Dolling in the chapter 'Neutron Scattering' in one of the volumes *Rare Gas Solids* edited by Klein and Venables (1977).

Interpretation of experimental dispersion curves was usually made in terms of harmonic lattice dynamical models. It was realised from the earliest experiments on lead that anharmonic terms were present and observable; in the later measurements on the quantum solids helium, para-hydrogen and ortho-deuterium effects due to anharmonicity were particularly evident. For most crystals quantitative investigation of the higher-order anharmonic terms proved to be difficult. In the temperature and pressure regimes in which deviations from harmonic behaviour could be explored, the effects on the phonon lineshapes were usually near the limit of available resolutions. Nevertheless, there were successful attempts to measure anharmonic effects such as phonon frequency shifts and lifetimes in several materials and even to observe interference effects between the one-phonon and multiphonon processes.

The extension of coherent, inelastic neutron scattering techniques to molecular solids in the early 1970s introduced significantly more complication, for molecules possess three rotational degrees of freedom and the symmetry of molecular solids is often rather low. However, the first molecular solid for which dispersion curves were measured at Chalk River was the high-symmetry cubic crystal hexamethylenetetramine. Data were obtained for all the high-symmetry directions and successfully analysed in terms of both force constants and interatomic potentials. However, detailed measurements on more complex molecular solids, especially biological systems, remained a major challenge.

The Recent Past

In the last decade neutron scattering has completed the transition from a novel research measurement to a powerful, widely used standard research technique, although, due to the high cost of neutron sources, the availability of neutron scattering is more restricted than that of X-ray scattering. In lattice dynamics (as in other applications of neutron scattering) the tendency is to investigate materials which are 'more difficult' in some sense. Solids with lower symmetry, more atoms or molecules in the unit cell, smaller samples of new and exotic materials, more extreme environments (particularly pressure), materials which show more subtle (or a greater variety of) transitions, and materials which are disordered in some sense, have all been more extensively explored in the last few years. However, despite the bewildering proliferation of materials on which lattice dynamical measurements have been made, we can identify several general classes of materials or types of measurement which have been of particular interest and have thus been assigned large periods of spectrometer time in various neutron laboratories.

Investigations of superconducting properties, manifested through the electron–phonon interaction as anomalies in the phonon dispersion curves, have been extensively pursued. The initial phonon measurements on superconducting metals such as Pb, the Pb–Tl–Bi alloy system and Nb and Mo and their alloys, have been extended to other transition-metal alloy systems and to the hexagonal close-packed transition-metal superconductors. The dispersion curves of several transition-metal carbides and nitrides were measured at Oak Ridge and the anomalous features observed in those which become superconductors were ascribed to a strong electron–phonon interaction leading to superconductivity. Many of these dispersion curves were interpreted in terms of sophisticated shell models. The highest superconducting transition temperatures occur in the cubic A-15 compounds, some of which undergo a structural phase transition near the superconducting transition. The suggestion has been made that the structural transition is due to a soft mode driven by an electronic instability related to the superconducting transition, and efforts have been made to measure the dispersion curves in these compounds. The available single crystals are generally very small, but results have been obtained for the archetypal compound. Nb_3Sn. Measurements of natural phonon linewidths as a function of temperature in both Nb_3Sn and in pure Nb clearly showed the effects of the gap energy on the width. Such changes provide a direct measure of the electron–phonon interactions, but are at the limit of present day experimental resolutions.

Materials which not only have great practical importance but also illustrate the fundamental physics of proton diffusion are the metal–hydrogen systems. These systems, in which large amounts of interstitial hydrogen are dissolved in the metal (usually a transition metal), have been extensively studied, particularly at the National Bureau of Standards. The archetypal metal hydride is PdH_x, which remarkably becomes supercon-ducting whereas pure Pd metal does not and even more remarkably the T_c for PdD is higher than that for PdH—in contrast to the normal isotope effect. This effect has been explained in terms of anharmonicity. Extensive measurements of dispersion curves in the deuteride $PdD_{0.63}$ have shown very large dispersion in the optic branches. Dispersion curve measure-ments have also been made in NbD_x for various values of x and in this case the optic modes show little dispersion, while the acoustic modes have a more complex behaviour than in PdD_x. Measurements have also been made in the rare-earth dihydride $CeD_{2.12}$. As more dynamical measure-ments become available it will be possible to test the theories of formation of metal–hydrogen alloys more thoroughly.

Materials with high ionic conductivity, the superionic or fast-ion conduc-tors, have attracted great interest because of their technological import-ance. The copper halides which have anomalously high ionic conductivity are close to the ionicity phase boundary between the zinc blende and the sodium or caesium chloride structures. Measurements of their disperson curves have been made with particular effort expended in investigations of the observed lineshapes to study the pronounced anharmonic effects. Many fluorite structures have superionic properties at high temperatures and extensive neutron measurements of the temperature dependence of phonons in several halides with the fluorite structure have been made at Harwell, Risø and ILL. Similar measurements have been made on the oxygen fluorite UO_2. These data have been used to test theories for the mechanism of the ionic conductivity based on anion disorder. Other materials which exhibit fast-ion conductivity are the silver halides in their high-temperature phase and other salts based on these halides. The temperature dependence of the phonons has been measured in several of these salts (particularly AgI) in an effort to correlate mode softening and apparent anharmonic effects with the onset of ionic conductivity. The β-alumina superionic conductors are rather different, for they show no well defined phase transition and the mobile ion is constrained to move in two dimensions by the layer structure. The low-frequency parts of the disper-sion curves have been measured for both Na^+ and Ag^+ β-alumina. Recently, efforts to understand the occurrence of fast ion conductivity in salts such as $LiSO_4$ have been made. In such salts the SO_4 ions are thought to be orientationally disordered and to aid in the Li diffusion. Lattice dynamical measurements would be of great help in confirming this prop-osed mechanism but single crystals are very difficult to grow.

Following the early measurements of Dolling and Brockhouse on pyro-lytic graphite in 1962 little further work on layer structures was done for several years. However, increased interest in the anisotropic properties of such structures and the increasing availability of single crystals (usually small) has spurred a large number of experiments in the last few years. The classic layer structure, graphite, was re-examined in 1972 and dispersion curves in several transition-metal dichalcogenides and monochalcogenides have been measured. Since these compounds all have rather complex structures with high-frequency optic modes, complete dispersion curves were not measured for any of them. Other more ionically bonded layer structures such as PbI_2, have also been examined. The 2H polytypes of both $TaSe_2$ and $NbSe_2$ aroused particular interest because both are metallic and, at low temperatures, exist in an incommensurate phase. The transitions appear to be due to the softening of an LA mode driven by a charge-density-wave instability. However, the situation is more complex, for in $TaSe_2$ at least there appears to be a 'lock-in' transformation at lower temperatures. The small size of available crystals has limited detailed lattice dynamical measurements which might elucidate the transformation mechanism. A field which has expanded rapidly in the last few years is that of intercalation compounds, whereby guest atoms or molecules are intercalated between the 'layers' of a host layer structure. The favoured host to date has been graphite (graphite intercalation compounds) but others are now being studied. Depending on the layer structure, the intercalant, its concentration and the conditions under which intercalation is carried out, a bewildering variety of properties and structures may result. Lattice dynamical measurements on some of these compounds are just beginning and will very likely expand in the future.

The lattice dynamics of molecular solids assumed increasing importance as single crystals became available. The generally lower symmetry and the larger number of degrees of freedom in these solids forces a heavy reliance on model calculations to interpret the measurements; present instrumental resolutions are rarely adequate to resolve experimentally closely spaced branches. If a model is not available, then the phonons can often be unambiguously determined only at higher symmetry points, as in CS_2 for example. However, if reliable interatomic potentials are known then very complex dispersion curves such as those in naphthalene can be measured. The librational degrees of freedom present in a molecular solid allow the existence of an orientationally disordered state in which the orientation of the molecules is disordered although their centres of mass lie on an ordered lattice. There have been several measurements of the collective excitations in such solids. The classic examples are the ammonium halides in which the orientation of the ammonium ion is disordered, and recently there has been much work done on KCN, β-N_2 and SF_6.

The earlier measurements in crystals containing heavy mass defects e.g. Cu(Au) by Svensson *et al*, or light mass defects, e.g. in Cu(Al) by Nicklow *et al*, were followed by studies of many other mixed systems such as high-concentration substitutional metal alloys, mixed diatomic compounds, mixed alkali halides and mixed ammonium halides. Mixed compounds of the halides were particularly popular since the halogens often substitute easily for each other. Moreover, a halogen may also be replaced by a cyanide ion which introduces another degree of freedom into the problem. The mixed system KBr:KCN, which shows a rather complex phase diagram, has recently been the subject of intense study. Low concentrations of molecular impurities in metals have also been investigated. The host phonons may then interact with the internal vibrations of the impurity. These experiments often involve the measurement of a small perturbation of the host phonons or an accurate measurement of lineshape in the defected crystal and are then at the limit of present day resolutions.

The fascination which phase transitions have exercised on condensed-matter scientists is as great as ever and the role which neutron scattering has played in elucidating details of the many types of transitions has steadily increased. More recently great interest has been generated by the discovery of phase transitions into incommensurate phases. These are phases in which two (or more) mutually incompatible elements of translational symmetry are present. The transition seems to occur as the result of a charge-density-wave instability and, since this is coupled to the phonons, it can be viewed as a soft-mode displacive transition. For metals it appears as a giant Kohn anomaly. The nesting of the Fermi surface necessary for such a transition is much more probable for lower-dimensional systems and the appearance of incommensurate phases in quasi-two-dimensional layer structures has already been mentioned. However, the most spectacular observations have been made in quasi-one-dimensional metals where the transition is a Peierls transformation. The linear platinum complex KCP and the one-dimensional organic metal TTF-TCNQ are examples in which such transitions occur and dispersion curves have been measured in these and related materials. Despite first appearances, these transitions do not follow exactly the pattern of a simple soft-mode instability. The most detailed investigations have probably been made on the insulator K_2SeO_4, and this material does appear to exhibit simple soft-mode behaviour. Another unusual material with an incommensurate phase is $Hg_{3-\delta}AsF_6$ in which the Hg ions form one-dimensional chains. Detailed dispersion curve measurements have been made on crystals of this material. In the incommensurate phase two new types of excitations are predicted to appear, amplitudons and gapless excitations termed phasons. The latter excitations are reported to have been observed in the incommensurate insulator biphenyl. Detailed measurements of the dynamics of these exotic materials have been hampered by the difficulties of obtaining suitable single crystals, but since an increasing number of materials is found to show such transitions, clearly the field will expand in the future.

Neutron scattering is usually considered a probe for bulk matter, but during the last decade a large effort has been put into its use as a surface probe. To compensate for the weak scattering from surfaces, neutron experiments use substrates with very large surface areas. Most of the structural and all the dynamical measurements have been made with graphite substrates. The structural studies of simple molecules, such as the rare gases, adsorbed on graphite, reveal highly complex phase diagrams as a function of temperature and coverage. Phase transitions of order–disorder, solid–liquid and commensurate–incommensurate character have been observed. Even more complex phases arise if more complex molecules such as ethane or pentane are absorbed. Dispersion curve measurements of the overlayer species are at the limit of present day neutron fluxes but have been made for ^{36}Ar and for H_2, D_2 adsorbed on grafoil. It is clear that detailed measurements of the lattice dynamics would be of great help in understanding the mechanisms responsible for this bewildering variety of phases.

The Future

Inelastic neutron scattering has always been intensity limited, consequently many phenomena for which it would be a suitable experimental probe cannot at present be investigated in detail. As new, more intense, sources are built (both reactor-based and accelerator-based) we may expect that some of these intensity limitations will diminish. The improved resolution allowed by higher intensity sources should enable natural phonon linewidths and lineshapes to be determined with much greater precision. The systematic study of phonon interactions with other excitations in the solid, e.g. other phonons, electrons, charge fluctuations, static or dynamic defects, vibronic excitations etc will thus become possible. The dynamical behaviour of a solid is specified by a knowledge of both the phonon frequencies and the phonon eigenvectors. The former are commonly measured by inelastic neutron scattering, but measurements of the latter are rarely made since accurate intensity measurements at several momentum transfers are required. Increased intensity will make measurement of the eigenvectors much more feasible and, when combined with the increased detail in the dispersion curves allowed by higher resolution, will enable more rigorous tests to be made of models for interatomic and intermolecular forces. With increased intensity it should also be possible to investigate the time-dependent lattice dynamics of metastable phases or systems which have finite relaxation times following a change in some environmental parameter or following the creation of a non-equilibrium phonon distribution.

Many of the current studies of excitations in partially disordered systems suffer from insufficient intensity and since investigation of these systems will assume increasing importance in the future, higher-intensity sources will be imperative. Inelastic scattering measurements of the dynamics of systems such as one-dimensional metals, orientationally disordered solids, intercalated systems, stretch-oriented solids and gas overlayers on surfaces will all benefit greatly from increased intensity. Further, measurements of the diffuse scattering, which is present in many of these systems, will also become more feasible.

A field in which inelastic neutron scattering has had only limited impact to date is the study of the dynamics of biological systems. These systems are usually complex, often partially disordered, sample sizes are small and specimens often deteriorate with time. Increased intensity will clearly alleviate many of these difficulties. However, a further severe problem with biological samples is the presence of intense incoherent scattering from hydrogen atoms. Separation of the coherent scattering from this spin-incoherent hydrogen background is possible by means of polarisation analysis, but makes heavy demands on intensity. With more intense sources, however, it may be the key to the effective application of neutron scattering to the dynamics of biological systems.

In the three decades since the first inelastic neutron scattering experiments were carried out lattice dynamical dispersion curves have been measured for a wide range of crystalline solids. This body of data has had an enormous influence on our understanding and description of interatomic and intermolecular forces in solids and, in several instances, has led to the development of new concepts to interpret the experimental observations. The technique is now recognised as the most powerful method for investigations of the lattice dynamics of solids. There is no doubt that its use as a probe of interactions on the microscopic scale will continue to expand in the future at both reactor-based and accelerator-based sources.

References

Bilz H and Kress W 1979 *Phonon Dispersion Relations in Insulators* (Berlin: Springer)

Egelstaff P A (ed) 1965 *Thermal Neutron Scattering* (London: Academic)

Horton G K and Maradudin A A (eds) 1974 *Dynamical Properties of Solids* vol 1 Crystalline Solids, Fundamentals (Amsterdam: North-Holland)

IAEA 1961 *Inelastic Scattering of Neutrons in Solid and Liquids. Proc. Symp. on Inelastic Scattering of Neutrons in Solids and Liquids* (Vienna: IAEA)

Klein M L and Venables J A (eds) 1977 *Rare Gas Solids* vol II (London: Academic)

7.4 Neutrons in Biology

B Jacrot

European Molecular Biology Laboratory, Grenoble, France

The tremendous success of X-ray crystallography in the period 1953–65 for the understanding of biological processes at the molecular level has quite naturally stimulated various scientists to investigate the usefulness of neutron diffraction in molecular biology. The complementarity of X-rays and neutrons in the study of organic compounds has been well known since the late 1950s and it was obviously interesting to use this complementarity to get further structural information on biological objects. I have in my file notes on discussions held at Saclay in 1964 in which experiments on protein and even virus crystals were suggested. The first effective investigation was done at Harwell in the UK. In 1965 Dorothy Hodgkin and Terry Willis started an analysis on vitamin B12, using X-ray and neutron diffraction in parallel, and the first report on their work was soon published in *Nature* (Moore *et al* 1967). Vitamin B12 is not a protein but is, in fact, a big organic molecule (molecular weight 1300) and analysis of the data proved that with neutron diffraction at 1.3 Å resolution an extremely clear map could be obtained, showing hydrogen atoms as negative contours and revealing the various chemical groups of the molecule. The *Nature* article had a strong impact and stimulated the pioneering work on neutron protein crystallography of Benno Schoenborn, who published his first results on myoglobin, a protein of molecular weight 16000, in 1969.

Protein crystallography is not the only structural approach in molecular biology. In the 1960s considerable interest was expressed in biological circles towards the aim of understanding the structure of complex systems, in particular biological membranes, muscles and viruses. The membranes, which limit the volume of a cell, and through which the cells communicate with the outside, are certainly not accessible as such to direct high resolution crystallography, but rather to methods giving much poorer information, such as the determination of the average electron density profile obtained from the Fourier transform of an X-ray diffraction pattern given by a sample made of parallel sheets of membrane. Acknowledging the possible complementarity of X-rays and neutrons, Parsons and Akers (1969) put a sciatic nerve in a 4 Å neutron beam at the high-flux reactor at Brookhaven. To a good approximation, the scattering by such a nerve is given by its membrane (myelin sheath) which is folded around the nerve in such a way that it provides a natural sample of ordered membrane sheets. The sample was measured, soaked in H_2O and in D_2O. Spectacular differences were observed, but not really analysed. The experiment was repeated by Kirschner and Caspar (1972) who developed an interpretation based on the contrast between the surrounding water (H_2O or D_2O) and the various chemical components of the membrane.

At the same time a group at Jülich started experiments on small-angle scattering from protein solutions. The first work published (Schneider *et al* 1969) was on haemoglobin and simply showed the possibility of using neutron beams, and it was not until 1972 (Schelten *et al*) that the same group did experiments with solvents made with various amounts of D_2O and showed that the square root of the intensity at the origin had a linear variation with the percentage of D_2O in the solvent, and that with haemoglobin in solution this intensity vanished for 40.5% D_2O. These results were easily understandable, as the scattering at zero angle has an intensity

$$I(0) = (\Sigma b - p_{sol} V)^2 N$$

where Σb is the sum of the scattering amplitudes of all atoms of a molecule, N is the number of molecules in the beam, V the volume occupied by each molecule, or more precisely the volume of the sample which is not occupied by the solvent. p_{sol} is the average scattering density of that solvent. So, by varying the relative amount of D_2O in the solvent it is possible to modify the relative contribution of the various chemicals constituting a biological object: protein, nucleic acids, lipids and sugar. The method of contrast variation which makes systematic use of that property was indeed well known in light and X-ray scattering. Neutrons, however, have a clear advantage over X-rays. In the latter case the scattering density of the solvent can be varied only by adding small soluble molecules such as sugar or salts. The range of density is limited, and, for instance, it is not possible to match with such a solvent the density of a nucleic acid. With neutrons, this is easily done with about 70% of D_2O in the solvent. Moreover, D_2O has very little (if any) effect on the structure of a biological macromolecule whereas salts may perturb strongly that structure, and are not uniformly distributed in the solvent. A systematic investigation of the method was made by Stuhrmann and published in 1974.

The method of contrast variation with neutrons is based on the very large differences in the scattering properties of H and D atoms. Another method of using that difference was proposed by Engelman and Moore (1972). They proposed to establish the architecture of a complex biological object, the ribosome, by replacing two proteins by their deuterated counterpart in a subunit of this ribosome, made of 20 different proteins and a piece of nucleic acid. Then the comparison between the scattering curves given by the native object and that with the two substituted proteins yields the distance between these two proteins. This method, known as the triangulation method, has been and is still widely used, in particular for the original problem for which it was suggested, namely the architecture of the ribosome (Moore 1979).

After a timid start, in about 1965 in various places, all the methods to be applied later were finally established in 1972. Since then those methods have been used, very often in combination, to solve problems in structural molecular biology. An example of the association of methods is low resolution crystallography. The concept of cancelling the contribution of a chemical component should strictly be valid only at the limit of zero angle. However, it remains a good approximation up to the resolution where the scattering or the diffraction can still be calculated from the local average scattering density and does not require knowledge of the position of individual atoms. So, diffraction from a single crystal soaked in buffer with a variable amount of D_2O, up to a resolution of about 15 Å, can be analysed to give rough structural information on each component of a complex particle, e.g. a virus.

It is interesting to follow the development of the various lines of research whose initial stages we have outlined. Two meetings held at Brookhaven National Research Laboratory in 1975 and 1982 produced useful records (Schoenborn 1976, 1984) which help in following the progress in the various areas.

Protein Crystallography

Neutron beams are weak compared with X-ray sources, and the work in this domain has been limited to crystals of molecules of low molecular weight—around 10 000—and which can be obtained in large sizes (a few mm^3). In all cases the structure of the protein had previously been observed by X-ray crystallography, and neutrons are used to acquire more information, essentially on the position of hydrogen atoms—for instance those involved in hydrogen (or deuterium) bonds or those of bound water molecules (see, for example the work of Hanson and Schoenborn (1981) on myoglobin). Some protons are especially interesting as they play a key role in the functioning of the enzyme. Kossiakoff and Spencer (1981) have been able to resolve a much debated issue on the chemical basis of the action of the protease trypsin.

Neutron diffraction is often done with crystals soaked in D_2O to mini-mise the background due to H atoms. So, it is possible to determine which H have been exchanged for D, providing information on protein dynamics (Kossiakoff 1982), which can be compared with that obtained by NMR from the same protein in solution (Bentley *et al* 1983).

Now, neutron protein crystallography is considered to be the final state of refinement of the analysis of a protein structure. The limitation of the application of the method is a problem of intensity. Today, if one has a high-flux reactor and a single detector, data on lysozyme can be collected to 2 Å resolution within 9 days (Lehmann, private communication); with a position sensitive detector it should be possible to have useful data on small and medium size proteins in a few days. This should lead to a much more systematic use of the method. Moreover, with fully deuterated proteins, as they can now be obtained by genetic manipulation in bacteria grown in D_2O, the range of applications should extend to larger proteins.

Contrast Variations from Solution Scattering

This method is easy to use and requires neither large amounts of material nor long periods of experimental time. For this reason it has been used by many in experiments on both high and medium flux reactors. Usually data are collected from many solutions with various amounts of D_2O, although it has been demonstrated by Stuhrmann (1974) that all this information is reducible to only three independent curves. In any case, the redundant information contributes to an improvement of the result of the analysis. The main difficulty in this analysis is that one is trying to obtain a three-dimensional structure from only one-dimensional data. This difficulty already existed in the analysis of small-angle X-ray scattering. A lot of work has been done, mostly by Stuhrmann, but also by Glatter and Luzzati to try and cope with that insoluble problem (see a review by Zaccai and Jacrot 1983) and in practice the analysis remains very empirical. With spherical objects which can be analysed with a one-dimensional distribution the difficulty does not exist, and this is why a lot of work has been done on viruses which are often approximated by a sphere, and contrast variation provides a radial distribution of the viral components (nucleic acids, proteins and lipids). This approach has been applied to many viruses during the last ten years.

A development of the method has been the determination of molecular weights, which is facilitated by the internal calibration of the beam and of the geometry given by the almost pure incoherent scattering of H_2O (Jacrot and Zaccai 1981). A special case of this molecular weight determination is the observation of complex formation between molecules. If those molecules are of a different chemical nature (e.g. nucleic acid and protein) it is possible to determine, using solvent with various amounts of D_2O, the stoichiometry of the various components in the complexes. This method has contributed to clarification in problems such as the interaction between transfer RNA and the specific enzyme which activates this molecule before its involvement in protein biosynthesis (Dessen *et al* 1978).

We described the method of triangulation developed to study the architecture of ribosome. Progress has been made recently in the use of contrast variation on a carefully reconstituted ribosomal subunit to separate the scattering of a single protein from that of the unit made to have a uniform scattering density matched with the appropriate solvent (see May *et al* in Schoenborn (1984)).

Neutron scattering from solutions is a developing method. It is being used more and more by biochemists to get rough structural information. One of its merits is that it can be done with rather small reactors, so the method is accessible to many users.

Low Resolution Crystallography

This method overcomes the main difficulty of small-angle scattering in that it provides three-dimensional information. Developed at Grenoble since 1976, the method found a very useful application in solving the structure of the nucleosome, the building unit of chromosome. Made of DNA and protein, this particle can be crystallised, but crystals diffract only up to 6.7 Å. With X-ray crystallography it is very difficult, if not impossible, to distinguish nucleic acid from protein. So, the neutron diffraction (Finch *et al* 1980) which gave independent maps of the DNA (from the data in 40% D_2O) and of the histone core (from the data in 70% D_2O) was most useful for the interpretation of the X-ray map. Unfortunately, data collection is slow, and so far the application of the method has been limited by the lack of dedicated instruments. Otherwise, it is likely to be of great importance for all the biological objects which are not amenable to high resolution X-ray crystallography. Recently it has been used to investigate the folding of the nucleic acid inside spherical viruses (G Bentley, D Wild and P Timmins, private communication), information which goes beyond the radial distribution deduced from solution scattering. In the future the method should be of importance for various problems, such as the structure of the ribosome (which has now been crystallised) or membrane proteins.

Membrane Structure

The development of research in this field has also been based on specific deuteration and H_2O–D_2O exchange. A review of the early work done in this area has been given by Worcester (1976), a pioneer in the field. Experiments done on artificial membranes with a deuterium label at different positions along the lipid molecule have provided an original method of determining the mean conformation of those molecules (Buldt *et al* 1979).

The natural membrane which has been investigated the most has been the purple membrane of *Halobacterium halobium*. This membrane, found in bacteria living in a very salty medium, is a two-dimensional crystal. Neutron diffraction with the natural membrane, or with the membrane labelled chemically or biochemically on a specific point, has largely contributed to the determination of its structure (see Jubb *et al* 1984). This is a good example of the complementarity between neutrons, X-rays, electron microscopy and biochemistry.

Dynamical Studies

The study of dynamics is one of the most important contributions that neutron scattering has made to physics and chemistry. Although biological macromolecules are far more complex than the material usually investigated, it was very tempting to apply neutron inelastic scattering to the study of protein and nucleic acid dynamics. The first work done was a study of the motion of water molecules in contact with DNA (Dahlborg and Rupprecht 1971). However, since that time, the emphasis has been mostly on protein dynamics (see Middendorf 1984). Obviously, neutron scattering is a low resolution method in view of the huge number of modes expected from the complexity of the molecule, and, *a priori*, it was feared that the measured spectrum would be that calculated from the rather simple models of protein dynamics (McCammon 1984) smeared with experimental resolution. In fact, the experimental spectrum is totally different but this negative result is actually an important one as it shows the necessity of using more realistic models to describe protein dynamics.

The perspectives in this area may be large. So far, analysis has been based on the incoherent scattering given by all the protons of the protein. It is quite realistic to imagine that, in the future, work will be done on deuterated proteins (in solution with D_2O) with only a few protonated residues which will dominate the incoherent scattering. Conversely, work on oriented samples (for instance membrane, or DNA) could help to isolate specific modes propagating in a well defined direction. Here again, a deuterated sample would be required to minimise the incoherent scattering.

Conclusions

This brief historical survey was not intended to be a review article. Many important contributions have been omitted. I have in mind, for instance, the fundamental instrumental development done at Grenoble by Ibel or at Brookhaven by Nunes. The future is very open, and there are now many biochemists familiar with neutron methodology, which has found its place in structural biology and in biochemistry, alongside much older techniques.

(a)

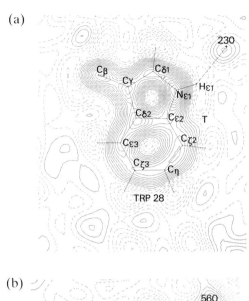

(b)

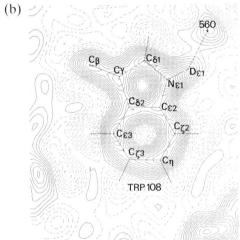

Figure 7.10 The structure of triclinic hen eggwhite lysozyme. Labile hydrogen atoms in a protein crystal may be exchanged for deuterium from a heavy water buffer. These labile sites are easily distinguished in high resolution neutron diffraction experiments by virtue of the large difference in scattering length ($b_H=-3.74\ F$, $b_D=6.67\ F$). In triclinic hen eggwhite lysozyme, Trp 28 is the tryptophan least accessible to the intracrystalline solvent; thus 28 $H_{\varepsilon 1}$ is seen here to be almost completely unexchanged (negative peak $H_{\varepsilon 1}$ (a)) after six weeks at pH 4.6 at room temperature. By contrast, the corresponding hydrogen in Trp 108 is fully exchanged (positive peak $D_{\varepsilon 1}$ (b)) because this residue is more accessible to the solvent. Note that both these hydrogens form hydrogen bonds to main chain carbonyl oxygen atoms. (Mason S A, Bentley G A and McIntyre G J 1984 in *Basic Life Sciences* vol 27 ed B P Schoenborn (New York: Plenum)).

Finally, in illustration, figures 7.10 and 7.11 present pairs of diagrams, for triclinic hen eggwhite lysozyme and satellite tobacco necrosis virus respectively, to give some idea of the detailed information on biological molecules which can now be obtained with neutrons.

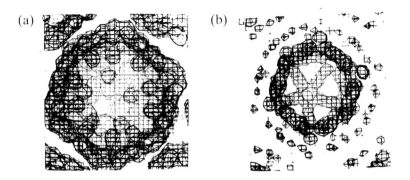

Figure 7.11 The structure of satellite tobacco necrosis virus (STNV) at 16 Å resolution using H_2O/D_2O contrast variation.

STNV is composed of 60 identical protein subunits and a single strand of RNA. The protein and RNA may be distinguished from each other at low resolution by matching the scattering power of the solvent regions within the crystal lattice to that of one of these two components by using a suitable H_2O/D_2O mixture. Thus, only the other component is seen. STNV has icosahedral symmetry and is approximately spherical in shape with a maximum diameter of 192 Å. The two views here show one half of the virus particle (i.e. approximately a hemisphere) as seen from its centre.

(a) At 68% D_2O, the solvent scattering is matched to the mean scattering of the RNA and only the protein, which forms the external shell of the virus, can be seen.

(b) At 40% D_2O, the mean scattering from the protein is matched out by the solvent and in this case the RNA is highlighted. The RNA is seen here to occupy a region just underneath the protein coat. The smaller features of density occurring in the protein region arise from inhomogeneities in the protein density, which are significant even at 16 Å resolution.

(Bentley G A, Lewit-Bentley A, Lilas L, Roth M, Skoglund U and Unge T manuscript in preparation.)

References

Bentley G A, Delepierre M, Dobson C M, Mason S A, Poulsen F M and Wedin R E 1983 *J. Mol. Biol.* **170** 243–7

Buldt G, Gally H U, Seelig J and Zaccai G 1979 *J. Mol. Biol.* **134** 673–91

Dahlborg V and Rupprecht A 1971 *Biopolymers* **10** 849–63

Dessen P, Blanquet S, Zaccai G and Jacrot B 1978 *J. Mol. Biol.* **126** 293–313

Engelman D M and Moore P B 1972 *Proc. Natl. Acad. Sci. USA* **69** 1997–9

Finch J T, Lewit-Bentley A, Bentley G A, Roth M and Timmins P A 1980 *Phil. Trans. R. Soc.* B **290** 635–8; see also Bentley G A, Finch J T and Lewit-Bentley A 1981 *J. Mol. Biol.* **145** 771–84

Hanson J C and Schoenborn B P 1981 *J. Mol. Biol.* **153** 117–46

Jacrot B and Zaccai G 1981 *Biopolymers* **20** 2413–26

Jubb J S, Worcester D L, Crespi H L and Zaccai G 1984 *EMBO J.* **3** 1455–61

Kirschner D A and Caspar D L D 1972 *Ann. NY Acad. Sci.* **195** 309–20

Kossiakoff A A 1982 *Nature* **296** 713–21

Kossiakoff A A and Spencer S A 1981 *Biochemistry* **20** 6462–74

McCammon J A 1984 *Rep. Prog. Phys.* **47** 1–46

Middendorf H D 1984 *Ann. Rev. Biophys. Bioeng.* **13** 425–51

Moore F M, Willis B T M and Hodgkin D C 1967 *Nature* **214** 130–3

Moore P B 1979 *Ribosomes; Structure Function and Genetics* ed G Chambliss *et al* (Baltimore: University Park Press)

Parsons D F and Akers C K 1969 *Science* **165** 1016–18

Schelten J, Schlecht P, Schmatz W and Mayer A 1972 *J. Biol. Chem.* **247** 5436–41

Schneider R, Mayer A, Schmatz W, Kaiser B and Scherm R 1969 *J. Mol. Biol.* **41** 231–5

Schoenborn B P 1969 *Nature* **224** 143–6

—— 1976 (ed) *Neutron Scattering for the Analysis of Biological Structures* (New York: Brookhaven National Laboratory)

—— 1984 (ed) *Neutrons in Biology* (New York: Plenum)

Stuhrmann H B 1974 *J. Appl. Crystallogr.* **7** 173–8

Worcester D L 1976 *Biological Membranes* vol 3 ed D Chapman and D F H Wallach (London: Academic) p. 1

Zaccai G and Jacrot B 1983 *Ann. Rev. Biophys. Bioeng.* **12** 139–57

8 Changing Techniques

8.1 The Time-of-flight Diffraction Method—Reminiscences

B Buras

Risø National Laboratory, Denmark

The year of 1955 was the year of the first 'Atoms for Peace' Conference in Geneva and the beginning of a 'reactor boom'. Many countries decided to acquire nuclear research reactors and so did Poland. It was a 2 MW reactor with enriched uranium, moderated and cooled by light water and a thermal neutron flux of about 10^{13} neutrons cm^{-2} s^{-1}.

It was decided to place the reactor at Swierk near Warsaw, where the Institute of Nuclear Research had just been organised. At that time I was in charge of a small laboratory within the Institute and in view of the great challenge of the expected new research tool I decided to shift from semiconductor physics to neutron diffraction. Several colleagues joined me, and although we had no experience in this field, we nevertheless began construction of a double-axis crystal diffractometer. Therefore, during a visit to the United States in 1956, I took the opportunity of visiting Brookhaven National Laboratory where, among others, Donald Hughes and Harry Palevsky were doing neutron inelastic scattering experiments using the time-of-flight (TOF) method, and there I had my first direct encounter with this technique.

On my way back from the USA I visited George Bacon at Harwell and received from him useful advice concerning the design of our neutron diffractometer. His instrument was already working and both the visit to Harwell and Bacon's newly published book *Neutron Diffraction* played an important role in acquiring the necessary know-how in this new field. In 1958 the reactor at Swierk became operational and neutron diffraction research started.

In 1960 I took part in a small seminar in Paris devoted to the TOF technique, used at that time solely for neutron cross section measurements and inelastic scattering. At this seminar, Peter Egelstaff mentioned briefly that the TOF technique could also be used for elastic scattering and thus for structural studies of condensed matter. I learned later that this idea was also briefly mentioned by G R Ringo. R D Lowde also saw the white spectrum of wavelengths as a virtue rather than a somewhat unsatisfactory object for monochromatisation. However, no experiments had been done at that time.

The use of the TOF technique for diffraction studies intrigued me, and in 1963 with the help of my colleagues I began the preparations for an experiment. At that time a Fermi chopper was in operation at Swierk and we were able to use it together with a hundred-channel time analyser built up by electron tubes and telephone relays which often broke down. The neutron gas detector was home made and filled with non-enriched BF_3. Figure 8.1 shows schematically the experimental set-up.

The chopper was supplying short ($80 \mu s$) neutron pulses of a continuous wavelength distribution. The pulsed white beam was scattered from an aluminium powder sample under a fixed angle $2\theta_0$, and recorded by the neutron detector connected to the multichannel time analyser. Whenever the neutrons passed the slits of the Fermi chopper the time analyser was triggered and began to record the neutrons diffracted by the sample as a function of the time t of arrival at the detector. Knowing the time t and the total flight path l, we could calculate the neutron velocity v and thus the wavelength λ of the diffracted neutrons. The Bragg equation had just been rewritten in the following form

$$2d_H \sin \theta_0 = \lambda_H = \frac{h}{mv_H} = \frac{h}{ml} t_H$$

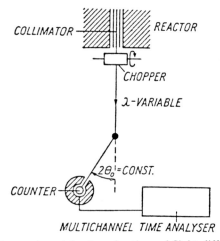

Figure 8.1 The first experimental set-up for time-of-flight diffraction studies.

where H stands for hkl, d_H is the interplanar spacing related to the hkl reflection, h is the Planck constant and m the neutron mass. Using this equation we could calculate for a given θ_0 and d_H the time t_H at which we should observe a peak in the intensity against time distribution. However, at the beginning we did not observe any peaks. Several days of battling with the electronics which broke down and improving the shielding to reduce the background from the other neutron beams did not help. I finally decided to stay overnight and measure under more favourable conditions, in particular as concerns electronic noise and the background (most of the shutters were closed at night). About four o'clock in the morning the peaks began to appear slowly. The number of counts per minute was about five and half of them were background. Once the peaks could be seen we were able within the next days to improve the results and send a note for publication (Buras and Leciejewicz 1963). It took us 40 hours to obtain the pattern shown in figure 8.2. The statistics were poor and so was the resolution ($\Delta d/d \sim 2.5 \times 10^{-2}$). Nevertheless we had experimental proof that the method does work. The formula for integrated intensity was derived at the same time and published in a separate paper (Buras 1963).

It was clear from the very beginning that the main advantages of the TOF powder-diffraction method were the simultaneous appearance of all reflections and the fixed geometry. The latter could be especially important for structural studies at high pressures. The fact that a high pressure cell could be constructed with only two windows—one for the incident beam and another for the diffracted beam—was appealing because it would make the construction of a cell much easier and thus enable high pressures to be reached. The first fixed geometry high pressure cell to be used in connection with the TOF powder method was built by R M Brugger et al (1967) at the Idaho Falls Material Testing Reactor.

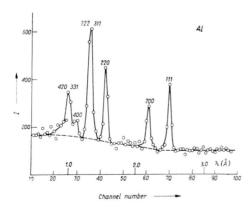

Figure 8.2 The first TOF diffraction pattern (Buras and Leciejewicz 1963).

As said above, in our experiments at the Swierk reactor, the statistics were poor. We needed higher intensity, although the losses in working only with short bursts of the main beam were partly compensated by recording a number of reflections simultaneously. In order to improve the statistics we took advantage of the fixed scattering angle and built a detector intercepting a large fraction of the diffraction cone. Nevertheless it was clear from the very beginning that steady state reactors— particularly those with a low or medium flux—are not the optimum sources for time-of-flight diffractometry. An inherent pulsed source would be ideal. Such a source existed within our reach—it was the nuclear pulsed reactor at the Joint Institute of Nuclear Research at Dubna, USSR. So, we went there. The reactor worked beautifully at an average power of 1 kW (it was increased later to 30 kW) with 5 pulses s^{-1} each about 100 μs long. The flux in the peak was not much higher than from our reactor at home, however we could use an order of magnitude longer flight path and larger samples. This resulted in a better resolution and shorter exposure times. However, Dubna at that time was not really prepared to host visiting scientists. Nowadays, a visiting scientist usually finds in a host laboratory technical help and a professionally built spectrometer. He also stays in a decent hotel or hostel. In the early 1960s at Dubna almost everything was supposed to be built from scratch; the technical help, especially for high precision items, was meagre and slow. The present comfortable Hotel Dubna did not exist and we had to stay in a kind of hostel in the Joliot–Curie street, two in a small room with iron beds and straw mattresses. In the tiny corridor was a sink with cold water. But, we had a stove which always heated the room more than adequately, a very important factor in Russian winters.

Despite these uninspiring conditions, and with the help of our Russian colleagues, we were able to build a primitive TOF spectrometer and take measurements. Their quality was much better than those obtained at Swierk. We presented them at the third 'Atoms for Peace' Conference in Geneva (1964), as a supporting argument for the need of intense neutron sources (Buras *et al* 1965a).

During the years 1963–8 work was going on at Swierk and Dubna, and in 1964 a TOF diffractometer was also installed at Risø National Laboratory, Denmark as a collaborative effort with the group at Swierk. During those early years the work was concerned in particular with instrumentation and methodological problems. Choppers with curved slits, diffractometers and detectors dedicated to TOF diffraction were built, the time focusing was studied, the resolution problem was analysed in detail and the derived formulae for resolution were checked experimentally. In 1965 the collaboration with Risø led to a test experiment that demonstrated that the TOF method could be used also to study diffraction from single crystals (Buras *et al* 1965b). This idea was later followed up at Argonne National

Laboratory (USA) and at Tohoku University (Japan) using two-dimensional and one-dimensional position sensitive detectors, respectively. A single-crystal TOF diffractometer is also working at Los Alamos National Laboratory (USA).

Our first two papers were published in the Polish journal *Nukleonika* and were hardly noticed, but the one presented at the Geneva Conference and several others published in international journals attracted some attention. However, a large number of scientists—among them also some colleagues in my own laboratory—did not believe that this method had a future. In a discussion at the seminar on Intense Neutron Sources (Santa Fe, 1966), where I presented the method, a respected scientist, while congratulating me on the results, said bluntly: this method is useless. Fortunately, not everybody shared his point of view and papers dealing with TOF diffraction began to appear from other countries as well. At steady-state reactors sophisticated choppers (among others the correlation chopper) were constructed. The main breakthrough for TOF diffraction came however when the technique was combined with neutrons produced by electron linear accelerators. When high-energy electrons (typically 100 MeV) bombarded a heavy target hard Bremstrahlung is produced which in turn, via the (γ, n) reaction, produces fast neutrons. The electron beam in a linear accelerator is pulsed and so is the fast neutron one. The time width of the pulse is of the order of $1 \mu s$, but the fast neutrons must be slowed down to the thermal energies necessary for diffraction studies and the final time width of the thermal neutron pulse is 50 to $100 \mu s$. The first TOF diffraction patterns obtained with the help of electron accelerators were obtained by M J Moore *et al* (1966) in the United States. M Kimura *et al* (1968) began to use the linear electron accelerator at Tohoku University, Japan for TOF diffractometry. About the same time, D H Day and R N Sinclair (1969) used the Harwell Linac for similar studies in the United Kingdom. The diffraction patterns obtained with the help of linear electron accelerators were characterised by an improved wavelength resolution and shorter exposure times when compared to TOF patterns at steady-state reactors.

By the end of the 1960s it became evident that more powerful pulsed sources were needed. The Joint Institute of Nuclear Research at Dubna, USSR decided to build a new pulsed reactor with an average power of several MW; it became critical in 1979 and is at present working at 2 MW. In the USA, the United Kingdom and Japan, another approach was favoured, namely to obtain pulsed neutrons via a spallation nuclear reaction. High energy protons (typically 800 MeV) bombarding heavy nuclei in the target cause evaporation of their components and for each high energy proton one obtains 20–30 fast neutrons. Again these neutrons must be moderated. But still the intensity of the pulsed thermal neutron beam produced can be made one to two orders of magnitude more intense than from linear accelerators.

The world's first prototype pulsed neutron spallation source was ZING-P at Argonne National Laboratory, put into operation in 1974†. Some years later (1977) the source at Los Alamos National Laboratory began to work, followed by sources dedicated to pulsed neutron research: the KENS source at the National Laboratory for High Energy Physics, Tsukuba, Japan (1980) and the Intense Pulsed Neutron Source (IPNS) at Argonne (1981). Very recently the dedicated Neutron Pulsed Source (ISIS) at the Rutherford–Appleton Laboratory produced the first neutron bursts (1984). These very intense pulsed neutron sources were built or are proposed not only for diffraction studies using the TOF method but also, among others, for inelastic scattering of neutrons and nuclear physics. However, the diffraction aspect has played and continues to play an important role when the decision to build them is made.

It has already been mentioned that the Institute of Nuclear Research at Swierk and Risø National Laboratory in Denmark collaborated in the field of development and use of TOF diffraction. This was one of the reasons why, when I left Poland in 1971, I decided to move permanently to Denmark. I was very much impressed by the double- and triple-axis spectrometers at Risø controlled at that time by PDP-8 computers or punched paper tape. However, the TOF diffractometer was already dismantled. It did not make sense to use the TOF method at a steady-state reactor, while inherent pulsed neutron sources were already in operation.

At that time I began to be interested in structural phase transformation studies under high pressure. The high pressure cell built by Brugger (1967), able to reach 40–50 kbars, was an attractive proposition. However, it required a fixed-scattering-angle diffractometer. In the absence of a TOF diffractometer Bente Lebech and I decided to use a triple-axis spectrometer. Keeping the scattering angle fixed we rotated stepwise the monochromator and by means of that we changed stepwise the wavelength of the incident monochromatic beam. The intensity of the diffracted beam was measured either directly by a neutron detector or reflected first by a single-crystal analyser and then recorded by a neutron detector. In the latter case, the rotation of the analyser followed the rotation of the monochromator accordingly. The reflections did not appear simultaneously, but the method turned out to be useful and was used in several phase transformation studies under high pressure.

Soon after, I became involved in synchrotron radiation research, and my previous involvement in the TOF technique had an interesting spin-off effect. In Poland, at the laboratory which I headed until 1968, there was already a group making semiconductor detectors. These detectors, when

---

† Replaced by a more powerful prototype called ZING-P', which was completed in 1977 and operated until 1980.

connected to a multichannel pulse-height analyser, measured the spectral distribution of photons in the hard X-ray and gamma range. In 1968 it struck me that if, in the TOF method, one replaces the polychromatic neutron source, the neutron detector and the multichannel time analyser by a polychromatic X-ray source, a semiconductor detector and a multi-channel pulse-height analyser respectively, then one has working fixed-scattering-angle X-ray diffraction equipment. Using Bremsstrahlung from an X-ray tube as a source of 'white' radiation we tested this idea. At the same time a similar test was made by Giessen and Gordon. So the X-ray energy dispersive diffraction method was born.

After my arrival in Denmark, in parallel to the neutron research, I continued studies on and with the X-ray energy dispersive method at the University of Copenhagen. In the mid 1970s, I learned some details about synchrotron radiation emitted by electrons in the DESY synchrotron at Hamburg. It became immediately clear that synchrotron radiation with its high intensity and smooth spectrum is ideal for X-ray energy dispersive diffraction. In 1976, together with my colleagues in Copenhagen, I made a test experiment at DESY. The result was so attractive that, since that time, I have mainly used synchrotron radiation in my research, and in particular for phase transition studies at high pressures, taking advantage of the fixed geometry of energy dispersive diffraction and the high brightness of synchrotron radiation sources.

People never forget their first love and like to recall the time of great emotional involvement. So I followed, and still follow, with great interest the development of both the TOF method and pulsed neutron sources. It is always a great pleasure to see the interesting results and the fine high resolution TOF neutron diffraction patterns ($\Delta d/d \sim 3 \times 10^{-3}$), so different from those recorded in the early days.

These reminiscences are neither a review nor the history of the time-of-flight diffraction method. They try only to give some glimpses of the time of its birth and development and the role it played in the development of X-ray energy dispersive diffraction. It should also be clear that many more scientists have contributed to its success than are mentioned in the text and references†. Without their work these reminiscences would not meet the requirements for inclusion in this book.

Acknowledgment

The author would like to thank all colleagues who have helped to collect the information necessary for writing these reminiscences. At the same time the author wishes to apologise for the fact that only a very small part

† A comprehensive discussion of TOF diffractometry can be found in Windsor (1981).

of the information he received was used, due to the limited length of this
note which is mainly supposed to cover the early days of TOF diffraction.

References

Brugger R M, Bennion R B, Worlton T G and Peterson E R 1967 *Research
    Applications of Nuclear Pulsed Systems* (Vienna: IAEA) p 33
Buras B 1963 *Nukleonika* **8** 259
Buras B and Leciejewicz J 1963 *Nukleonika* **8** 75
Buras B, Leciejewicz J, Nitc W, Sosnowska I, Sosnowski J and Shapiro F L 1965b
    Proc. UN Int. Conf. on the Peaceful Uses of Atomic Energy, 1964 (New York:
    United Nations) VII pp 447–54
Buras B, Mikke K, Lebech B and Leciejewicz J 1965a *Phys. Stat. Solidi* **11** 563–73
Day D H and Sinclair R N 1969 *Nucl. Instrum. Methods* **72** 237
Kimura M, Sugawara M, Oyamada M, Yamda Y, Tomiyoshi S, Suzuki T,
    Watanabe T and Takeda S 1968 *Laboratory of Nuclear Science, Tohoku Uni-
    versity, Sendai, Japan, Report TUEL-5*
Moore M J, Kasper J S and Menzel J H 1966 *Nature* **219** 848
Windsor C G 1981 *Pulsed Neutron Scattering* (London: Taylor and Francis)

8.2   From BF$_3$ Counter to PSD. An Impressive and Continuous Increase of the Data Acquisition Rate

P Convert and P Chieux

Institut Laue–Langevin, Grenoble, France

To look at neutron scattering from the detector end is certainly looking at
it through a distorting lens, and this is even more the case if we do it only
from experience gained in diffractometry in Grenoble. Nevertheless it is
certainly worth making a brief journey through the exciting years of
instrumental development we have just traversed, and which still continue.
When we started our experimental work in the mid 1960s the prehistoric
times of the 'stop-watch and crankhandle' spectrometer for powder diffraction measurements were gone, as well as the more convenient, but rather
qualitative, continuous angular scans with chart-recording of the intensities
measured by a ratemeter. Already most neutron diffractometers were
automated for angular step scans, and had a neutron monitoring device

(generally a fission detector) to correct for the instabilities of the neutron sources. Neutron detectors were commercially available in an almost standard form of $^{10}BF_3$-filled cylinders of 1–2 inches diameter and 5–10 inches long (Reuter Stokes, USA; 20th Century, GB: Le Materiel Telephonique (LTM), France) which were set coaxial with the scattered beam to maximise the detection efficiency. A system of cadmium Soller-slits was installed at various places along the neutron beam, in particular in front of the detector, to improve the angular resolution. This period was also characterised by the appearance of the first transistorised electronics which, with the digital encoders, were essential in the automation of the goniometer displacements and progressively reduced the size and price of the neutron-counting chains.

However, although in the late 1960s rather good spectrometers were available with a fair resolution, sufficient stability, convenient automation and punched paper data output compatible with computer analysis, the main drawbacks remained the lack of neutron flux and the inefficiencies of the methods of data acquisition. At the time when dedicated high-flux reactors became available for research, the idea of investigating new detection devices was beginning to occur. A rather simple step taken to improve the detection efficiency was to install a bank of identical detectors with their own Soller collimators, and this was tried in many places (e.g., the Curran diffractometer at Harwell). However, and under the influence of the progress in nuclear physics detectors, it was quite challenging to investigate the possibility of constructing neutron position-sensitive detectors (PSDS) with adequate spatial resolution.

Two localisation principles were tried, one based on resistance–capacitance (RC) position encoding along a highly resistive wire lying parallel to the scattering vector (Borkowsky, Kopp (Oak Ridge) 1968, Kjems (Risø) 1970), the other based on a multi-electrode localisation system (Allemand, Jacobe, Roudaut (Grenoble) 1968 patent). At that time the production of neutron detectors entered what we might call the industrial level. This became especially apparent in Grenoble where the conjunction of an old neutron research team (Roudaut et al, CENG), a semi-industrial electronics laboratory (Allemand et al, at the Laboratoire d'Electronique et de Technologie de l'Informatique (LETI) of CENG) and the impulse of the creation of a new breed of neutron research instruments (Maier-Leibnitz et al at the ILL) allowed the specification and construction of two large-scale $^{10}BF_3$ PSDS.

A linear PSD of 400 cells covering 80° (2θ), with a radius of curvature 1.5 m and an effective detection height of 75 mm, was available at the Siloe reactor of the CENG in Grenoble in 1970. It was filled with $^{10}BF_3$, at one atmosphere, which, with a detection gap of 15 mm ensured a detection efficiency of 9.5% at a wavelength of 1.29 Å. It worked by direct charge collection and used a matrix system of electrodes reducing the number of amplifiers to 40. This first design revealed immediately the advantages of PSDs in simultaneity, speed and stability (<1%), but the simple geometry of direct charge collection gave a poor signal-to-noise ratio (~2) for the neutron detection. A second design working in the proportional mode, with an efficiency of 44% at $\lambda = 2.4$ Å, and a stability of about 0.1% was operational in 1975 on the D1B diffractometer (figure 8.3) at the ILL. This PSD, built in collaboration (CENG, ILL, LMT) was the first of a series of 'Banana Detectors' commercialised by LMT, now LCC, Bollene, France, which progressively changed the methods of collecting powder-diffraction data, as we shall see later.

The second type of large $^{10}BF_3$ gas PSD built by LETI in the early 1970s was a two-dimensional multielectrode 64×64 cells of 1×1 cm^2 each, for the world-famous small-angle neutron scattering (SANS) machine D11 at ILL. Good collaboration between the scientific team (Springer *et al*) and the different technical services ensured a detector very well adapted to the huge dimensions of the instrument (up to 80 m long) as well as to the available system of data acquisition and evaluation. Since D11 was based on the use of cold neutrons, a high detection efficiency was easily achieved. Immediately, in 1973, the live display (figure 8.4) in a pseudo-tri-dimensional (isometric) representation of the small-angle scattering pattern was a breakthrough. Of course, here also, the direct charge collection was replaced by the proportional mode in 1975. More elaborate versions of the PSD such as one with 128×128 cells of 5×5 mm^2 were designed and commercialised by LETI. But the basic characteristics of the D11 instrument and detector remained a standard for many similar instruments installed on research reactors all over the world.

Figure 8.3 The 'banana' detector at D1B in 1974, with Pierre Convert at the teleprinter.

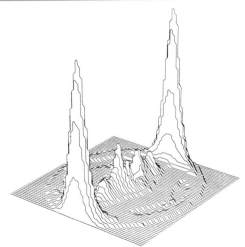

Figure 8.4 Isometric representation of the three-dimensional small-angle scattering pattern of frog's sciatic nerve (D Worcester and K Ibel).

In the meantime, the RC encoding technique was being developed in the USA with ^{3}He gas filling. In 1975 a cylindrical one-dimensional PSD of length 500 mm with a resistive stainless steel wire capable of resolving 250 spatial elements was available at Oak Ridge for diffusive magnetic scattering (Cable) and later for studies of disordered materials (Narten). Two-dimensional PSDs of size 20×20 cm^2 were also constructed at the same period at both Oak Ridge (Kopp *et al*) and Brookhaven (Radeka *et al*). Schoenborn at BNL defined the optimal configuration for the use of such two-dimensional PSDs for crystallographic studies of biological samples which were so time-consuming. The use of high gas pressure (6 atm of ^{3}He, 4 atm Ar, 0.5 atm CO_2) achieved a resolution of 3 mm and an efficiency of 70% at 2.8 Å. This, in principle, opened the field of protein crystallography to PSDS.

In the late 1970s and early 1980s, several teams concerned with the development of existing neutron pulsed sources or the creation of powerful new machines (ISIS, England, SNQ, Germany) reconsidered the possibility of using scintillators for neutron detection and this on the basis of two main characteristics: their response time which is 10 times faster than the gas detection process, and their small thickness (1–2 mm) well adapted to time-of-flight experiments. Davidson and Wroe at the Rutherford Appleton Laboratory worked on a juxtaposed mosaic of glass scintillator elements with a special method of fibre optic coding to reduce the number of photomultipliers. Another localisation principle working with a continuous area of scintillator is to allow the light produced to disperse before entering an array of photomultipliers whose analogue output signals are used to derive the position of the incident neutron. This system, called the

Anger camera, was very successfully optimised by Strauss *et al* at Argonne (1981) largely due to the arrangement of the camera optics, with a thin air gap between the scintillator plate and the light disperser. A large camera with a 30×30 cm^2 ^{6}Li glass scintillator coupled to an array of 7×7 square photomultiplier tubes (51×51 mm^2) was incorporated in a single-crystal diffractometer at IPNS and gave the Laue pattern shown in figure 8.5. Although the result obtained, especially the 2.7 mm resolution, was quite satisfactory, many improvements were foreseen. In particular, as noted by Schelten *et al* (Germany) who thoroughly developed an Anger camera and used it for powder diffraction studies, a considerable gain was expected from the production of a scintillator with a short emission time, yet a high light output. Research on the scintillator material as well as on the decoding system is being pursued, a resolution of about 1.2 mm being now achieved. Clearly this approach to neutron detection has not yet achieved its full performance and several years will be needed before a balanced view of the advantages and disadvantages of scintillators compared with gas detectors can be given.

Coming back to the gas PSDs, the late 1970s and early 1980s were marked in the USA by the production and commercialisation (Borkowsky, Kopp) of several types of large size, 65×65 cm^2, two-dimensional PSDS for SANS, all on the RC encoding principle. At the same time, significant progress was made by Radeka *et al* at Brookhaven in the optimisation of the gas mixtures, e.g. (^{3}He+C$_3$H$_8$) allowing a position resolution close to 1 mm, high neutron detection efficiency and very good neutron–gamma ray discrimination. In Europe, the large one-dimensional and two-dimensional ^{10}BF$_3$ PSDS were installed at many reactors, the latest version of the LETI

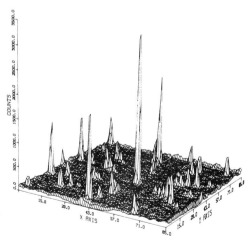

Figure 8.5 Laue pattern for crystal of K$_{0.26}$WO$_6$. The intensities are summed over a wavelength range of 1.0–3.0 Å.

banana detector being one with 800 cells (80°(2θ), pressure 2 bars) and a spatial resolution of 2.6 mm, available in 1980.

Between 1980 and 1985, the ILL 'Second Souffle' produced a renewed effort in instrumentation. Focusing monochromators were installed on many instruments, generally multiplying the neutron flux at the sample by a factor of 3. We give in figure 8.6 a picture of the D7 composite focusing monochromator. A complementary effort was of course made on the detection side. Small PSDs were developed (Jacobe *et al*) either one-dimensional (64 cells) or two-dimensional (16×64 cells) with a detection area of 8×16 cm², filled with ³He mixtures. These PSDs had very good stability (better than 0.03%), and because of their high gas pressure (up to 15 bars for hot-source instruments), good resolution (2.5 mm or 2.5×5 mm²) and very high efficiency (~85%). When installed on high flux diffractometers they made it possible for the first time to work at 0.1% statistical accuracy. However, for very good resolution ($\Delta d/d$ down to 0.05%) a powder diffractometer was still designed (D2B, Hewat) on the

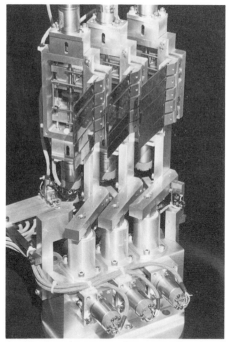

Figure 8.6 A doubly focusing monochromator for the diffractometer D7 at the Institut Laue–Langevin. There are three independent crystal systems which can be adjusted to give horizontal focusing. Vertical focusing is achieved by a lever system attached to the crystal supports. Focusing conditions vary with Bragg angle, and hence with the neutron wavelength, and are optimised by the instrument computer (A K Freund).

principle of a large bank of 64 single ^{3}He detectors equipped with 5 minute Soller collimators separated by 2.5°.

In the field of crystallographic studies 'fly's eye' cameras, i.e. small PSDs for following a single Bragg peak, were introduced. But the main development, namely a large ^{3}He gas PSD of 512×16 cells covering 64°×4° with a resolution of 2.5×5 mm^2 for protein crystallographic studies, is only now being completed, due to the considerable effort required not only on the hardware side but also on the development of reliable algorithms to treat on-line the huge amount of data produced. Other PSDs for this type of study, and their associated software, are under development (e.g. DB21 at ILL) and with them, as well as with some pulsed source instruments with on-line data reduction, we are entering a new era. Data treatment itself is entering a semi-industrial level requiring team-work of engineers, analysts and physicists for its conception. The concept of the 'raw data', as we used to call it, is altered. The ordinary scientists will see the outcome of their experiments through the glasses of the software package.

Although we have tried so far to keep this brief survey of the development of neutron detectors at a narrow technical level, it is clear that the progress made was concurrent with significant developments in other fields such as electronics or 'informatics'† as well as with the general evolution of neutron scattering techniques for research. Electronics from valves to germanium, and then silicon, transistors, integrated circuits, dedicated integrated circuits and microprocessors—this is a well known story. It may be more relevant here to remember that around 1965, for a single ^{10}BF$_3$ detector of roughly one pint in volume, the associated electronics (preamplifier, amplifier, discriminator and trigger output) occupied ten times that volume, and the scaler itself was twice as large again. In 1985 a multielectrode linear PSD which, with one amplifier per cell, is a very big consumer of electronics, houses all this electronics, including discriminators and logic, within a volume no larger than the PSD box itself. This, roughly speaking, for one counting chain is a gain of two orders of magnitude in volume, with much better performances and reliability. It is accompanied also by a gain of two orders of magnitude in price. At the same time the dead-time, corresponding to the black-out after one neutron detection, has gone down from 10 μs to 1–2 μs, i.e. reaching the physical limit given by the duration of the detection process in the gas which is dominated by the collection time of the electrons. Still one more order of magnitude might be gained on this value with the scintillators, being again limited by the physics of the detection process. In every respect electronics is becoming invisible in the detection chain.

---

† 'L'informatique'; the whole range of equipment used for data-handling, computation, manipulation and display.

Informatics, a latecomer to neutron scattering, moved at a slower pace from the rearguard of data treatment to the front line of the measurements where it took some time to replace existing equipment and modify experimental practices. This progression, intimately connected with increases in memory size, speed, peripherals, and the lowering of computer prices, has not been an easy matter. The borderlines between electronics and informatics were for a long time a main concern for instrument 'responsibles' i.e. those scientists charged with the care of individual instruments. And the introduction and development of elaborate software on the experimental site was often a matter of debate amongst scientists, since it required a consensus on data acquisition and data-handling procedures and also because it modified the practices of reserving the experimental time and effort for the acquisition of good raw data (i.e. neutron counts from the hardware counting chains). It is interesting here to see briefly what happened at service institutions like the ILL where every gain in neutron flux or in the speed of data acquisition has a tendency to reduce the average duration of an experiment; and every progress in machine automation or in the availability of data-reduction packages makes neutron scattering more appealing to categories of users less professionally trained. As soon as the spectrometers started operation (1973), priority was given to installing an automatic procedure for storage and safeguard of formatted spectra of raw data in the DEC 10 main computer. This proved particularly useful for the high data density of PSDs or time-of-flight machines. A second step (1976–80) was the on-line software display of the data for proper decision making. This required, when possible, extension of the storage capacity on the instrument to accommodate data for a complete experiment. Then, well documented software packages were developed for on-line rough evaluation of properly corrected spectra, again with the idea of a more efficient use of the neutron beam time. Automation of various sample environment facilities (sample changers, temperature controllers, etc) was a relief to the experimentalist. Finally, the software is in specific cases becoming an integral part of the data acquisition chain, algorithms being developed to find out the significant information, the rest being dropped. To illustrate this huge evolution which took place at various places let us again consider what happened at ILL on a few diffractometers equipped with PSDs.

The small-angle neutron scattering instrument D11 which opened a new instrumental field was from the beginning very well designed for convenient use with a hardware display of the last two spectra in various forms, including an isometric representation. However, with the fast data acquisition rate allowed by the 4096 cell PSD it very soon became customary to run quick and interactive experiments of 10 to 30 minutes per spectrum and to do so 24 hours a day with a rapid turnover of experimentalists of different backgrounds. Therefore a need developed for elaborate but self-exploratory on-line procedures to compare spectra, to fit crude laws to the data, etc. In 1976 the D11 software package of Ghosh *et al* accompanied by a storage capacity of several hundred spectra on the dedicated PDP 11/23 computer and a Tektronix display with hardcopy, answered this problem so well that it sometimes led users to neglect a careful *a posteriori* data analysis. Of course, later developments in the instrument automation (changes in wavelength and in the sample-detector distance) as well as in the control of the sample environment (temperature, sample changers, etc) were again added for the convenience of the users, but without basically improving the instrument performance.

The value of the banana PSD of D1B in the well established field of powder diffractometry took much longer to be recognised. The advantages of the stationary detector, covering continuously and with high stability a large angular range, were progressively discovered. It became a very powerful tool for measuring small sample modifications such as weak lines of magnetic structures (e.g. $CeAl_2$, 1980), physisorbed monolayers, etc and this was the occasion for considerable progress in the provision of accurate sample environments. On the other hand, fast data acquisition, correction and display as offered by a dedicated computer was only achieved in the early 1980s, a significant delay compared with the situation of the small-angle instrument. Repetitive short scans, immediately processed and displayed for a detailed study of a sample evolution with temperature, for example, could then routinely be obtained as shown in figure 8.7. With the gain in flux due to the focusing monochromators and to the mounting of PSDS on high-flux beam holes, real-time kinetic experiments with a time resolution down to a few seconds for irreversible phenomena (e.g. chemical reaction) and a fraction of a millisecond for reproducible phenomena (using stroboscopic methods) are now being developed.

Even with a huge effort in semi-automatic ways of data processing, we reach a point where data output from the machines might become too large to be cross checked by the experimentalist. Automatic on-line data reduction has to be introduced. This is the case for the large PSDS on single-crystal diffractometers such as D19, equipped with a 8192 cell two-dimensional PSD for protein crystallographic studies which will routinely

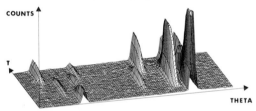

Figure 8.7 Powder diffractograms showing the α–β–γ sequence of changes in $LiIO_3$ as a function of temperature.

produce a spectrum or frame every 10 seconds, and this continuously for days. Algorithms for the real time search, characterisation and integration of Bragg peaks have been developed. This is a huge step which has been made. Although the experimentalist's practical approach has always been a highly educated and skilfully trained one, part of his know-how will now be built into the computer software. More than ever, creativity and discovery will be made with the help of, but with the requirements seen through the eyes of, the whole technical and scientific community.

As we have seen, in order to gain efficiency and variety in the use of neutron beams, considerable progress has been made in instrumentation and this has been not only on the hardware side of neutron production and detection, but also, by necessity, on data-handling software routines. In every respect, the neutron instruments and even rather simple diffractometers have become highly elaborate machines requiring the team effort of professionals for their design, although the progress of informatics has permitted considerable simplification of a lot of time-consuming, repetitive and even delicate tasks and made access to the machines easy for the non-specialist. A neutron experiment now generally interacts highly with measurements by other techniques and also, to a large extent, with model calculations. Efforts are always made to shorten the time scale of these interactions, but here we are generally dependent on personal relationships. Although we have restricted our brief survey to PSD spectrometers, the high specificity of the present day instruments (in vector transfer and resolution range for example) often implies complementary measurements on various spectrometers, and a well balanced view of their capabilities. This depends considerably on human factors and often necessitates time and collective effort if it is to be achieved. In this respect the advent of new machines on pulsed neutron sources is certainly a challenge.

8.3 Neutron Spin Echo Spectroscopy

F Mezei

Hahn-Meitner Institut and Technische Universität, Berlin

The neutron spin echo (NSE) principle, a conceptually new approach to inelastic neutron scattering, was discovered in 1972—completely accidentally, of course. The following account retraces the line of thought which led to the discovery, and also explains the conceptual significance of NSE in the broader context of neuton scattering methods.

The idea of NSE spectroscopy was born one morning in April 1972 at the corner of Alagut and Attila streets in Budapest—while waiting at the traffic lights. By that time I had been working in neutron scattering for nearly four years at the 5 MW research reactor of the Central Research Institute for Physics in the hills on the outskirts of Budapest. In the first 18 months of my apprenticeship I came to the conclusion that without polarised neutrons nothing conclusive could be achieved in the study of magnetic alloys which I was supposed to be continuing. Thus—without giving much thought to the severe flux limitations of our reactor—I spent the rest of those years trying to put together an inelastic scattering spectrometer with polarisation analysis, i.e. the most flux demanding machine, which was at that time considered as impractical at even the highest flux reactors. To make things worse, I did this using odd pieces left over from previous pioneering work in magnetic chopping and correlation spectroscopy and by only exploring ways of improvement which were virtually without cost. My first publication on neutron scattering was yet to appear.

The day before that morning I tested a new version of the simplest and cheapest spin flipper device known at that time, the 'Kjeller 8' field flip coil (Abrahams et al 1962) and, to my great surprise, I observed a wiggle in the tuning curve of my set-up. Over dinner I convinced myself that this wiggle was a manifestation of Larmor spin precession. This struck me as an instant revelation that neutron spins in a beam behave according to classical mechanics, contrary to the common belief that only 'up' and 'down' spin-$\frac{1}{2}$ states are of physical relevance. (I remembered very well a saying which one of my professors at Eötvös University in Budapest attributed to Landau: 'Do it classically unless you are forced to use quantum mechanics!') It was in fact only several years later that I made the effort to take a rigorous look at this problem and I showed that, in the absence of appreciable Stern–Gerlach type beam-splitting, particles in a beam can indeed be described as point-like classical magnetic moments (Mezei 1980). This means there is a classical spin direction unit vector S following the $dS/dt=\gamma_L[S\times H]$ equation, where $\gamma_L=2.916\,\mathrm{kHz\,Oe^{-1}}$ and H is the magnetic field (in a magnetic medium it has to be replaced by B). This fact was common knowledge in the 1930s, but has since vanished from common thinking (Darwin 1927). So, while going about some business in town that morning, I kept pondering about what those simple classical Larmor precessions could be used for. First I came to the idea of the very simplest spin flipper, the flat-coil device, which is the key component in NSE technique (and which, in various combinations, has by now become the most common neutron spin flipper device), and then to the principle of NSE spectroscopy itself. In fact it was quite fortunate that at that time I was not aware of previous work on Larmor precessions in neutron beams (actually dating back to Frisch 1938), and therefore I did bother to spend some time meditating about the potential utilisation of this 'new' phenomenon.

The flat-coil spin flipper device only needs a few metres of insulated wire in order to be fabricated and a simple DC power supply in order to be operated. One of the two hastily made prototypes I tested with full success on the same day is shown in figure 8.8 and its operation is explained in figure 8.9. Assume a neutron enters the coil with its spin S parallel to the external field H. Inside the coil S starts to precess around the resulting field H', which is the sum of H and the field H_c produced by the coil. For example, if H' is bisecting the directions of H and H_c and the neutron spends time equal to half a precession inside the coil, the effect of the coil is to turn S from the direction of H into that of H_c (and vice versa). With properly chosen H and H_c any kind of spin rotation can be produced, e.g. $90°$ or $180°$ flip.

With a $90°$ flip, i.e. setting the spins (originally parallel to H) perpendicular to H, we can initiate Larmor precessions. In a homogeneous field the Larmor precession angle φ is given as

$$\varphi = \gamma_L H l / v$$

where l is the length of the neutron path and v the neutron velocity. Thus, if v is not strictly uniform in the beam, sooner or later the precessions get out of phase and become unobservable (see left-hand side of figure 8.10). There is, however, a simple trick (borrowed from the well known nuclear magnetic resonance spin echo method) to get rid of this trivial dephasing effect. It consists of reversing the sense of the precessions after a given distance l_0 (point B in figure 8.10) and making the precessions unwind. Thus at point C characterised by the condition $H_0 l_0 = H_1 l_1$ the total Larmor precession angle $\varphi = \varphi_0 - \varphi_1$ is zero for all neutrons, independently of velocity, i.e. the initial precessing polarisation at point A is recovered at C. Hence the name I adopted for the method.

The clue of using this simple echo phenomenon for inelastic scattering spectroscopy is the following. Let us put the scattering sample at point B in

Figure 8.8 The first step in NSE spectroscopy: an improvised flipper coil used in the very first test in April 1972.

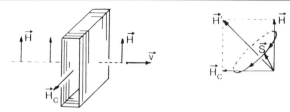

Figure 8.9 An example of Larmor spin precession inside the flat coil spin flipper device.

figure 8.10 and determine the precessing polarisation of the scattered beam at point C. If the neutron velocities change at B, the difference $\varphi = \varphi_0 - \varphi_1$ will deviate from zero, and actually (to first order) this deviation is a direct measure of the neutron energy change: $\varphi = \omega t$, where the proportionality constant is $t = \hbar \gamma_L l_0 H_0 / m v_0^3$. Thus the distribution of φs at the echo position C reproduces the ω energy variation of the scattering function $S(q, \omega)$. Therefore, taking into account that for small energies ($\omega \ll k_B T$) $S(q, \omega)$ is an even function of ω, the measured echo polarisation signal $P_{NSE} = \langle \cos \varphi \rangle$ actually corresponds to the Fourier transform of the scattering function,

$$\int S(q, \omega) \cos(\omega t)\, d\omega \simeq \int S(q, \omega) \exp(-i\omega t)\, d\omega = S(q, t)$$

where $S(q, t)$ is the time dependent correlation function describing the atomic dynamics of the sample, with t being the real physical time. For example for a simple relaxation process $S(q, t) \propto \exp(-\Gamma t)$, where Γ is the relaxation rate.

The real significance of the method is that—via the residual Larmor precession angle φ—it allows direct comparison of the initial and final velocities for each neutron individually. Thus, in contrast to the comparison of average beam energies measured in two separate steps in the

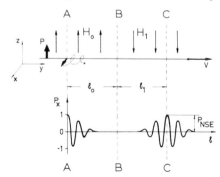

Figure 8.10 Dephasing of Larmor precessions in a beam and the spin echo rephasing effect.

conventional methods, in NSE we are concerned with a single-step direct difference measurement. Thus the monochromatisation (and hence intensity) of the neutron beam becomes decoupled from the inelastic resolution. For example 10^{-5} resolution in neutron velocity change is routinely achieved at 10% velocity spread in the beam. This decoupling is the fundamental novelty of NSE. We have to see this in the context that the increase of resolution in conventional neutron scattering is much more limited by the decrease of beam intensity to which it is coupled than by any other, more direct, experimental difficulty.

Up to this point we have considered the use of the method as it was originally formulated (Mezei 1972) for the investigation of small neutron energy changes (quasielastic scattering). The method, however, can be generalised to the high resolution investigation of large energy transfer processes (Mezei 1978), e.g. to phonon linewidth measurements. The principle of this kind of application is a long way from the simple echo considerations above, and fundamentally it boils down to the following. In conventional neutron scattering experiments we determine six parameters, i.e. the components of the ingoing and outgoing neutron momenta k_0 and k_1, respectively, and then we calculate the four parameters relevant to the sample scattering function $S(q, \omega)$, namely the momentum transfer vector $q = k_0 - k_1$, and the energy transfer $\hbar\omega = \hbar^2(k_0^2 - k_1^2)/2m$. In contrast, in NSE we probe the scattering function by observing the total Larmor precession angle $\varphi = \varphi_0 - \varphi_1$, which is itself a function of k_0 and k_1. What we have to achieve is that the function $\varphi(k_0, k_1)$ reduces (actually to first order around the (q, ω) spot of our interest) to a function of the $k_0 - k_1$ and $k_0^2 - k_1^2$ combinations of k_0 and k_1 only. Thus we probe the four-parameter function $S(q, \omega)$ with a matched four-parameter function $\varphi(q, \omega)$. This reduction of the relevant parameter set from (k_0, k_1) to (q, ω) is the general meaning of the echo condition, and it can be experimentally achieved in more complicated situations by adjusting not only the magnitude, but also the geometrical shape of the precession fields.

To conclude let us go back to the chronology. Before the month of April 1972 was over I was able to demonstrate at the Budapest reactor that the echo trick works. By the end of 1972 I had already shown at the Institut Laue–Langevin in Grenoble that the method has the predicted inelastic scattering capabilities (Mezei 1983). My proposal to build an NSE spectrometer IN11 was approved in early 1973 by the ILL council headed by Professor Mössbauer and the instrument was completed by 1977 in collaboration with John Hayter and Paul Dagleish (Dagleish *et al* 1980). The 2 m long precession field solenoids of this spectrometer (figure 8.11) can produce useful fields up to 750 Oe through which, for example, 8 Å wavelength neutrons undergo 55 000 radians precession. Hence we obtain the very high resolution representing the current ultimate limit in inelastic neutron scattering, as illustrated by the results in figure 8.12.

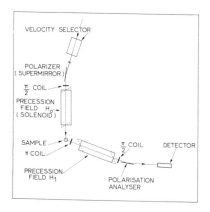

Figure 8.11 Schematic lay-out of IN11 NSE spectrometer at ILL, Grenoble. The dashed lines indicate the effective precession field geometry used to establish generalised echo conditions (see text).

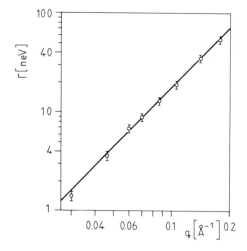

Figure 8.12 An experimental example; relaxation rates against momentum transfer measured in an immunoglobulin solution (Alpert *et al*). The neV energy resolution has been achieved with an incoming beam of 1.5 meV average energy and 0.5 meV energy spread.

References

Abrahams K, Steinsvoll O, Bongaarts P J M and de Lange P W 1962 *Rev. Sci. Instrum.* **33** 524

Alpert Y, Cser L, Farago B, Franek F, Mezei F and Ostanevich Y M 1985 *Biopolymers* **24** 1769

Dagleish P A, Hayter J B and Mezei F 1980 *Neutron Spin Echo* ed F Mezei (Heidelberg: Springer) p 66

Darwin C G 1927 *Proc. R. Soc.* **117** 258
Frisch O R, von Halban H and Koch J 1938 *Phys. Rev.* **53** 719
Mezei F 1972 *Z. Phys.* **255** 146
—— 1978 *Neutron Inelastic Scattering* (Vienna: IAEA) p 125
—— 1980 *Imaging Processes and Coherence in Physics* ed M Schlenker *et al*
(Heidelberg: Springer) p 282
—— 1983 *Physica* **120 B** 51

8.4 Neutron Optics

H Rauch

Atominstitut der Österreichischen Universitäten, Wien, Austria

Introduction

The first diffraction experiments from crystals performed in 1936 by
Halban and Preiswerk and by Mitchell and Powers† were the beginning of
neutron optics. The related diffraction phenomena create the basis of
chemical and magnetic structure analysis which became an important
discipline of condensed matter research. Many of these aspects have been
discussed in the previous chapters. Here we present a narrower view of
neutron optics where the microscopic structure of matter recedes into the
background and the refraction and diffraction from macroscopic objects
come to the forefront. All these phenomena are associated with the
wave-like properties of neutrons characterised by their de Broglie
wavelength $\lambda = h/mv$. There are striking analogies to experiments with light
and, particularly, to X-ray optics but also to electron and ion optics.

The theoretical description is based on the Schrödinger equation

$$\Delta\psi + (2m/\hbar^2)(E - V(r))\psi = 0. \tag{1}$$

The most dominant neutron–nucleus interaction occurs in the s state and
can be written in the form of the Fermi potential with the low energy
scattering length b as the one and only parameter

$$V(r) = (2\pi\hbar^2/m)b\delta(r). \tag{2}$$

†These papers appear in Chapter 2.

In the case of a collective interaction with many scattering centres with a particle density N, a mean interaction potential $\overline{V}$ is defined

$$\overline{V} = \frac{2\pi\hbar^2}{m} b_c N \tag{3}$$

where the mean phase shift defines the related coherent scattering length b_c. Energy conservation directly yields the index of refraction which in the case of low absorption takes the form (Goldberger and Seitz 1947)

$$n = (1 - \overline{V}/E)^{1/2} \simeq 1 - \lambda^2 N b_c/2\pi. \tag{4}$$

This is the fundamental quantity of standard neutron optics. In the case of magnetic interaction a term $\pm\mu B$ has to be added to $\overline{V}$ where μ is the magnetic moment of the neutron and B is the magnetic induction.

Many fundamental quantum mechanical propositions can be tested on a macroscopic scale by using neutrons which are massive particles with a distinct internal quark structure (u–d–d) and which are subject to strong, electromagnetic, weak and gravitational interaction. This creates a strong motivation for neutron optical experiments. Only a few examples can be discussed in this chapter and for details the reader is referred to more extensive reviews (Sears 1982, Klein and Werner 1983, Rauch 1986).

Instrumentation Aspects

The greatest impact on advanced neutron instrumentation derives from the development of neutron guides (Christ and Springer 1962, Maier-Leibnitz and Springer 1963). These systems transfer, by total reflection inside polished hollow tubes, the luminosity existing near the source to many instruments placed at low background positions far from the source. The divergency of these beams is determined by the critical angle for total reflection which follows from equation (4)

$$\sin \theta_c = (1 - n^2)^{1/2} = \lambda(N b_c/\pi)^{1/2} \simeq \theta_c \tag{5}$$

and which for thermal and cold neutrons is of the order of 10–60 min of arc. Supermirrors consisting of properly graduated layers can increase the angular region of total reflection by a factor of about three (Mezei and Dagleish 1977). They will be used for advanced neutron guides, polarisers and analysers, particularly in combination with high resolution spectroscopy. The gravity refractometer combines total reflection and the action of gravity on the neutron and is mainly used for precise scattering length measurements (Koester 1965).

The luminosity of neutron sources is always orders of magnitude less than that of light, X-ray or electron sources. This challenge resulted in the development of sophisticated beam-tailoring components. Neutron guides and supermirrors are two of them and neutron spin echo systems are another tool for a most efficient use of the available neutron flux (Mezei 1972, 1980). Each neutron carries its own clock in the form of its Larmor rotation angle and small energy changes of the order of neV become observable. Charged particle physics has gained substantially by the invention of phase space cooling; no similar methods for neutrons are known at present but various bunching systems, particularly for pulsed sources, are feasible by mechanical and electromagnetic means (Maier-Leibnitz 1966, Buras and Kjems 1973, Steyerl 1975, Rauch 1985).

Diffraction from Macroscopic Objects

The formal analogy between the Helmholtz equation in optics and the time-independent Schrödinger equation for matter waves results in similar diffraction phenomena for both kinds of radiation. The diffraction from macroscopic objects is described by the coherent superposition of Huygens waves. In the case of small phase shifts between the emitted waves the Fraunhofer diffraction regime dominates, whereas for large phase shifts Fresnel diffraction becomes dominant. Single-slit diffraction was first observed by Maier-Leibnitz and Springer (1962) and has been investigated in detail by Landkammer (1966) and by Shull (1969) who used the narrow rocking curves of perfect crystals arranged in a non-dispersive position (figure 8.13). The observed broadening due to a slit of width d agrees with the well known Fraunhofer formula

$$I(\theta) \propto \frac{\sin^2(\pi d \sin \theta/\lambda)}{(\pi d \sin \theta/\lambda)^2}. \qquad (6)$$

An even higher angular sensitivity can be achieved by using the central peak of multiple Laue-rocking curves whose widths are of the order of the ratio of the lattice constant d_{hkl} divided by the thickness of the reflecting crystal ($d_{hkl}/t \sim 0.001$ sec of arc; Bonse et al 1979a). The diffraction of 1.8 Å neutrons from slits with widths up to 5 mm has been observed by this method (Rauch et al 1983). The observed broadening can also be understood as a result of the Heisenberg uncertainty principle caused by the position determination inherently caused by the slit.

The diffraction from ruled gratings has been observed for thermal (Kurz and Rauch 1969), cold (Graf et al 1979) and ultra-cold neutrons (Scheckenhofer and Steyerl 1977). The classical diffraction experiments from a double slit, an absorbing wire and from a Fresnel lens have been

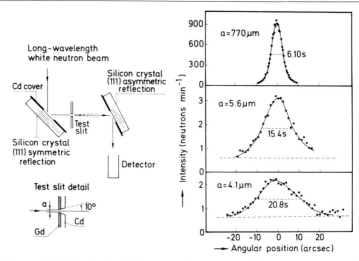

Figure 8.13 Single-slit diffraction of neutrons with a wavelength of 4.43 Å observed with a non-dispersive double perfect-crystal arrangement (Shull 1969).

performed in close analogy to experiments with optical light by using a 10 m long optical bench. Characteristic results are summarised in figure 8.14 (Klein *et al* 1981a,b Gaehler *et al* 1981, Zeilinger *et al* 1981). These measurements belong to the Fresnel diffraction regime and the results are

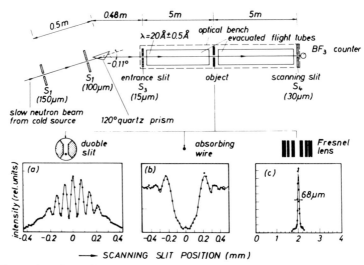

Figure 8.14 Experimental set up for classical diffraction experiments from macroscopic objects. (*a*) double-slit diffraction (Zeilinger *et al* 1981), (*b*) diffraction at an absorbing wire (Gaehler *et al* 1981), (*c*) diffraction from a Fresnel zone plate (Klein *et al* 1981b).

in good agreement with the theory, to an extent which permits the extraction of new upper limits for non-linear terms in the Schrödinger equation (Gaehler *et al* 1981).

Neutron Interferometry

The first attempt at neutron interferometry by Maier-Leibnitz and Springer (1962) was based on wave-front division and biprism deflection. First-order interferences have been observed but due to the small beam separation (~50 μm) the coherence was destroyed when a sample was introduced into the coherent beams (Landkammer 1966, Friedrich and Heintz 1978). This situation was changed completely by the invention of perfect-crystal interferometers where widely separated beams are produced by dynamical diffraction phenomena from a monolithically designed perfect crystal. Figure 8.15 shows the first interference pattern observed at a 250 kW research reactor (Rauch *et al* 1974). High contrast interference patterns and interferences up to very high orders were later observed by this technique (e.g. Bonse and Rauch 1979b).

Many classical experiments on quantum mechanics have been accomplished on a macroscopic basis by perfect-crystal neutron interferometry. To this category belong the first verification of the 4π-symmetry of spinor wavefunctions which has been achieved independently by Rauch *et al* (1975) and by Werner *et al* (1975) (figure 8.16). The spinor wavefunction within a magnetic field propagates as

$$\psi(\alpha) = \exp(-iHt/\hbar)\psi(0) = \exp(-i\mu\boldsymbol{\sigma}Bt/\hbar)\psi(0) = \exp(-i\boldsymbol{\sigma}\alpha/2)\psi(0) \quad (7)$$

and, therefore,

$$\psi(2\pi) = -\psi(0) \qquad \text{and} \qquad \psi(4\pi) = \psi(0) \quad (8)$$

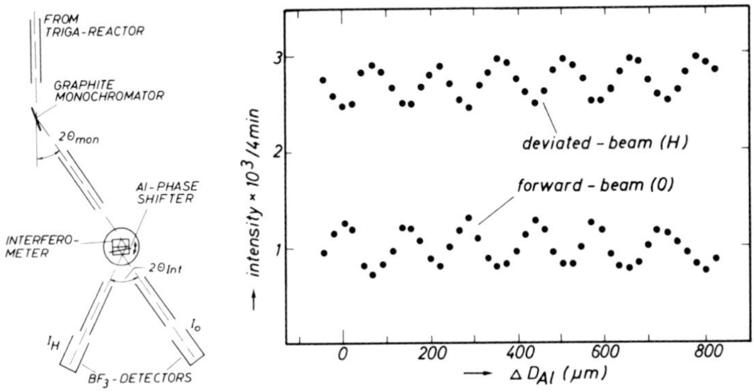

Figure 8.15 First perfect-crystal interference pattern observed at a 250 kW research reactor (Rauch *et al* 1974).

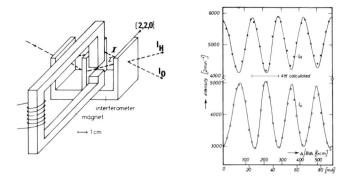

Figure 8.16 Verification of the 4π-symmetry of spinor wavefunctions by means of a perfect-crystal interferometer and an electromagnet (Rauch *et al* 1975).

where α is the Larmor precession angle around the magnetic field $\boldsymbol{B}$ and $\boldsymbol{\sigma}$ are the Pauli spin matrices. The interference pattern measured with unpolarised neutrons shows directly the typical 4π-symmetry when the magnetic field of the electromagnet is varied ($I \propto (1 + \cos \alpha/2)$). Later, these experiments were repeated with higher precision and have been extended by using polarised neutrons for a direct verification of the quantum mechanical spin superposition law (Summhammer *et al* 1983, Badurek *et al* 1983).

One of the most famous experiments with the neutron interferometer was the direct observation of the gravitationally induced phase shift on the neutron wavefunction (Colella *et al* 1975). This has been extended to the observation of the effect of the earth's rotation on the neutron phase (Staudenmann *et al* 1980). Figure 8.17 shows the result of this experiment where the interferometer was rotated around a vertical axis. The observed rotational Sagnac effect depends on the area enclosed by the coherent beams, on the colatitude angle of the place where the experiment is performed, on the earth's rotation frequency and on the orientation of the

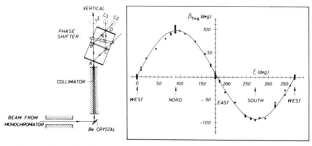

Figure 8.17 Gravitationally induced phase shift caused by earth's rotation when the perfect-crystal interferometer is rotated around a vertical axis (Staudenmann *et al* 1980).

interferometer relative to the earth's rotation axis, i.e. on the orientation of the interferometer plane relative to the compass points.

Ultracold Neutrons

They are defined as having kinetic energies smaller than the mean interaction potential ($E \leqslant \bar{V}$, equation (3)) and, therefore, they are totally reflected at all angles of incidence. Their energy is of the order of 10^{-7} eV and their wavelength is of the order of 1000 Å. Therefore, optical components can be used and gravitational and magnetic forces have strong influences on the beam paths.

The first experiments with these neutrons dealt with their production (Lushikov et al 1969, Steyerl 1969) and storage inside material containers where the storage times are limited by inelastic scattering processes at the surface. Storage times comparable to the natural lifetime of the neutron (11 min) have been observed within a superconducting toroidal hexapole storage ring (Paul and Trinks 1978, Kuegler et al 1978). Liouville's theorem and the small portion of ultracold neutrons in a Maxwellian spectrum limit the achievable density within various storage systems to values of the order of $1\,cm^{-3}$. Therefore, strong efforts have been made to increase the intensity either by Doppler shift at rotating totally reflecting turbine plates (Steyerl 1975, Kashukeev et al 1975) or by down scattering in liquid He-4 (Golub et al 1983).

There is increasing interest in bottled neutrons because the sensitivity of many experiments depends on the interaction time. Typical examples are

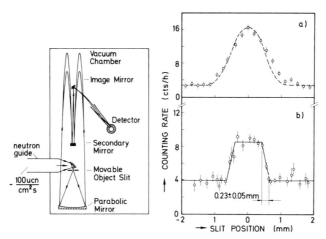

Figure 8.18 Scheme of the two-mirror neutron microscope and data scans for a 1.06 mm wide object slit for (a) a one-mirror system and (b) the nearly achromatic two-mirror system (Herrmann et al 1985).

investigations concerning the decay of the neutron, the search for an electric dipole moment or for a finite charge on the neutron. Many of these aspects have been discussed in detail in comprehensive reviews (Steyerl 1977, Golub and Pendlebury 1979). Recently the formation of resonances in synthetic potential wells and their energy splitting in the case of coupled macroscopic resonators has been observed (Steinhauser *et al* 1980, Steyerl *et al* 1981) and neutron microscopy with correction for gravitationally induced chromatic aberation has been introduced (Schütz *et al* 1980, Arzumanov *et al* 1984, Herrmann *et al* 1985). Figure 8.18 shows the principle of such an achromatic two-mirror microscope with gravity focusing and magnification of 50.

References

Arzumanov S S, Masalovich S V, Strepetov A N and Frank A I 1984 *JETP–Lett.* **39** 590

Badurek G, Rauch H and Summhammer J 1983 *Phys. Rev. Lett.* **51** 1015

Bonse U, Graeff W and Rauch H 1979a *Phys. Lett.* **69A** 420

Bonse U and Rauch H (eds) 1979b *Neutron Interferometry* (Oxford: Oxford University Press)

Buras B and Kjems J K 1973 *Nucl Instrum. Methods* **106** 461

Christ J and Springer T 1962 *Nukleonika* **4** 23

Colella R, Overhauser A W and Werner S A 1975 *Phys. Rev. Lett.* **34** 1472

Friedrich H and Heintz W 1978 *Z Phys.* B **31** 423

Gaehler R, Klein A G and Zellinger A 1981 *Phys. Rev.* A **23** 1611

Goldberger M L and Seitz F 1947 *Phys. Rev.* **71** 294

Golub R, Jewell C, Ageron P, Mampe W, Heckel B and Kilvington I 1983 *Z. Phys.* B **51** 187

Golub R and Pendlebury J M 1979 *Rep. Prog. Phys.* **42** 439

Graf A, Rauch H and Stern T 1979 *Atomkernenergie* **33** 298

Herrmann P, Steinhauser K A, Gaehler R, Steyerl A and Mampe W 1985 *Phys. Rev. Lett.* **54** 1969

Kashukeev N T, Stanev G A, Yaneva N B and Mircheva D S 1975 *Nucl. Instrum. Methods* **126** 43

Klein A G, Kearney P D, Opat G I and Gaehler R 1981b *Phys. Lett.* **83A** 71

Klein A G, Opat G I, Cimmino A, Zeilinger A, Treimer W and Gaehler R 1981a *Phys. Rev. Lett.* **46** 1551

Klein A G and Werner S A 1983 *Rep. Progr. Phys.* **46** 259

Koester L 1965 *Z Phys.* **182** 328

Kuegler K S, Paul W and Trinks V 1978 *Phys. Lett.* **72B** 422

Kurz H and Rauch H 1969 *Z Phys.* **220** 419

Landkammer F J 1966 *Z Phys.* **189** 113

Lushikov V I, Pokotilovsky Yu N, Strelkov A V and Shapiro F L 1969 *Sov. Phys.–JETP Lett.* **9** 23

Maier-Leibnitz H and Springer T 1962 *Z Phys.* **167** 386

—— 1963 *J. Nucl. Energy* A/B **17** 217

Maier-Leibnitz H 1966 *Sitzungsber. Bayr. Akad. Wiss* **16**

Mezei F 1972 *Z. Phys.* **255** 146

—— (ed) 1980 *Neutron Spin Echo (Lecture Notes in Physics* **128**) (Berlin: Springer)

Mezei F and Dagleish P A 1977 *Commun. Phys.* **2** 41

Paul W and Trinks V 1978 *Fundamental Physics with Reactor Neutrons and Neutrinos* (ed T von Egidy) (Inst. Phys. Conf. Ser. No 42) p. 18

Rauch H 1985 *Neutron Scattering in the Nineties* (Vienna: IAEA) p 35

—— 1986 *Contemp. Phys.* **27** 345

Rauch H, Kischko U, Petrascheck D and Bonse U 1983 *Z. Phys. B* **51** 11

Rauch H, Treimer W and Bonse U 1974 *Phys. Lett.* **47A** 369

Rauch H, Zeilinger A, Badurek G, Wilfing A, Bauspiess W and Bonse U 1975 *Phys. Lett.* **54A** 425

Scheckenhofer K A and Steyerl A 1977 *Phys. Rev. Lett.* **39** 1310

Schütz G, Steyerl A and Mampe 1980 *Phys. Rev. Lett.* **44** 1400

Sears V F 1982 *Phys. Rep.* **82** 1

Shull C G 1969 *Phys. Rev.* **179** 752

Staudenmann J L, Werner S A, Colella R and Overhauser A W 1980 *Phys. Rev. A* **21** 1419

Steinhauser K A, Steyerl A, Scheckenhofer H and Malik S S 1980 *Phys. Rev. Lett.* **44** 1306

Steyerl A 1969 *Phys. Lett.* **29B** 33

—— 1975 *Nucl Instrum. Methods* **125** 461

—— 1977 *Neutron Physics* (Springer Tracts in Modern Physics **80**) (Berlin: Springer) p 57

Steyerl A, Ebisawa T, Steinhauser K A and Utsuro M 1981 *Z. Phys. B* **41** 283

Summhammer J, Badurek G, Rauch H, Kischko U and Zeilinger A 1983 *Phys. Rev. A* **27** 2532

Werner S A, Colella R, Overhauser A W and Eagen C F 1975 *Phys. Rev. Lett.* **35** 1053

Zeilinger A, Gaehler R, Shull C G and Treimer W 1981 *Symp. Neutron Scattering, Argonne, AIP-Conf. Proc.* **89** 93

8.5   Neutron Diffraction Topography

J Baruchel† and M Schlenker‡

†Institut Laue-Langevin, Grenoble, France
‡Laboratoire Louis Néel du CNRS, Grenoble, France

Most imaging techniques are based on lenses: this is the case for the standard optics of light with which all of us are familiar (binoculars, cameras, microscopes) as well as for electron microscopy or ultrasonic microscopy. But lenses can only be made if the probe interacts strongly enough either with matter (light, ultrasound) or with easily manipulated fields or potentials (electrons), so that large deviations of rays can be produced; this is not the case for X-rays and neutrons.

One form of imaging has long been known, based on variations in absorption encountered in simple transmission, and this is radiography. However, the fact that both X-rays and, as this volume abundantly shows, thermal neutrons can be Bragg-diffracted by crystals provides the means for making images with considerably more information: the techniques by which this is done are called 'topographic' to emphasise the distinction with the standard X-ray or neutron diffraction approach.

These methods are based on two simple facts: first, the diffracted intensity depends on how perfect or imperfect the crystal is (this is often called extinction) and secondly, for neutrons but not for X-rays, the magnetic structure, i.e. the way magnetic moments are arranged on a microscopic scale, strongly influences the diffracting power.

In the classical approach to diffraction these effects are averaged out over the whole volume of the specimen by placing the detector far away from it. However, if a rather parallel and monochromatic incident beam is used, and if a position-sensitive detector is placed close to the specimen, the *local* variations in scattered intensity will show up as contrast, thus producing an image of the inhomogeneities, in terms both of lattice perfection (showing crystal defects) and of magnetic structure (showing the magnetic domains), in a single-crystal sample (figure 8.19).

Neutron-diffraction topography started in 1971. Very interesting results have been obtained in the observation of crystal defects, which can be performed non-destructively with quite bulky specimens. For example, Tomimitsu (1981, 1983) studied the texture of Bridgman-grown crystals of Cu–5% Ge, leading to a proposal for a new type of layered substructure. This accounted for both the striations visible on the surfaces of the crystals and those present on the topographs, and suggested a new type of growth mechanism.

In magnetism the technique has revealed completely new kinds of domains in antiferromagnetic materials, which just could not be observed by other methods. Examples are the details of the spin-density wave domains in antiferromagnetic chromium (Ando and Hosoya 1972, Davidson *et al* 1974) and the interpretation of domains in MnF_2 in terms of the piezomagnetic properties of the material (Baruchel *et al* 1980). In helimagnetic terbium it has been possible to distinguish between the chirality domains occasioned by right-handed and left-handed helices (Baruchel *et al* 1981), by using polarised neutrons: with suitable polarisation only one type of domain will diffract into each of a pair of magnetic satellites. A further example is illustrated by figure 8.20, which indicates how in the

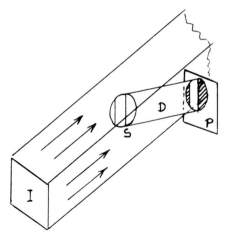

Figure 8.19 Principle of neutron diffraction topography. The single-crystal specimen S is immersed in a neutron beam I and set for a Bragg reflection, producing diffracted beam D. A neutron-sensitive photographic detector P placed across D near S will record variations in diffracted intensity, due, for example, to different magnetic domains in the sample, as contrast.

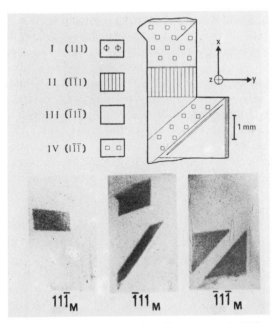

Figure 8.20 Neutron topographs of antiferromagnetic NiO made using magnetic (superstructure) reflections: only one type of domain is imaged on a given topograph, and it is then unambiguously characterised.

case of antiferromagnetic NiO, neutron topography can directly show, and unambiguously characterise, domains corresponding to different arrangements of magnetic moments (Baruchel *et al* 1977).

This technique, which is very simple in conception and in instrumental requirements, has the disadvantage of being slow and having rather poor resolution, but it offers some quite unique possibilities and has already produced some novel physical information.

References

Ando M and Hosoya S 1972 *Phys. Rev. Lett.* **29** 281
Baruchel J, Palmer S B and Schlenker M 1981 *J. Physique* **42** 1279
Baruchel J, Schlenker M and Barbara B 1980 *J. Mag. Magn. Mat.* **15–18** 1510
Baruchel J, Schlenker M and Roth W L 1977 *J. Appl. Phys.* **48** 5
Davidson J B and Case A G 1976 *Proc. Conf. on Neutron Scattering.* CONF-760601-P 2, Oak Ridge National Laboratory USA, p 1124
Davidson J B, Werner S A and Arrott A S 1974 *Am. Inst. Phys. Conf. Proc.* **18** 396
Hosoya S and Ando M 1978 *J. Appl. Phys.* **49** 6045
Schlenker M, Baruchel J, Perrier de la Bathie R and Wilson S A 1975 *J. Appl. Phys.* **46** 2845
Tomimitsu H 1981 *Phil. Mag.* A **43** 469
Tomimitsu H, Doi K and Kamada K 1983 *Physica* **120 B** 96

9 Some Views of The Future: Reactor and Accelerator

9.1 The Future with Accelerator-based Sources

G H Lander

Argonne National Laboratory, USA

Historical Background

Neutrons were produced by accelerating charged beams onto heavy metal targets even before the discovery of fission in 1939. As early as 1948 pioneering work was carried out with pulsed neutrons for making transmission measurements as a function of energy, but the rapid growth in both reactor fluxes and techniques for neutron scattering using monochromators in the 1950s and 1960s put pulsed sources at a disadvantage. Accelerator technology then began to gain on reactor advances for producing high intensity neutron beams and in the early 1970s both the Harwell electron linac and the first *proton* spallation source at the Argonne National Laboratory started to perform experiments in condensed matter research (Carpenter 1977). In the late 1970s three machines were under construction, the new HELIOS linac at Harwell, the KENS machine at Tsukuba, Japan and the Intense Pulsed Neutron Source (IPNS) at Argonne. Pulsed neutrons had come of age†, and new developments in accelerator technology promise to increase the flux available by a factor of perhaps 10^3. In the meantime, both the Rutherford and Los Alamos Laboratories realised that by modifying existing proton machines extremely useful pulsed spallation sources could be produced. Both of these sources started operating in 1985

---

† See, for example, Windsor (1981).

and by late 1986 were a factor of ten more intense then IPNS. The status of proton spallation sources is given in table 9.1.

Table 9.1 Proton spallation sources. This table gives the specification of present (and proposed) proton spallation sources in the world. The first such proton source operated at Argonne in 1974–5 with 0.1 μA current and an energy of 200 MeV. We have not covered electron driven sources here, the largest of which, HELIOS at Harwell, UK, also performs a good deal of condensed matter research.

Facility	Accelerator	Particle energy (MeV)	Time-average current (μA)	Average pulsing frequency (Hz)	Source pulse width (μs)	Target material	Status
ZING-P' Argonne, USA	Synchrotron	500	3	30	0.1	^{238}U	Operated 1977–80
WNR Los Alamos, USA	Linac	800	3.5	120	4.0	W	Started 1977
KENS-I KEK, Japan	Synchrotron	500	2	15	0.07	W	Started 1980
IPNS-I Argonne, USA	Synchrotron	500	12	30	0.1	^{238}U	Started 1981
SNS(ISIS) Rutherford, UK	Synchrotron	800	200	50	0.2	^{238}U	Started 1985
KENS-I' KEK, Japan	Synchrotron	500	10	15	0.5	^{238}U	Started 1985
WNR–PSR Los Alamos, USA	Linac + storage ring	800	100	12	0.27	^{238}U	Started 1986
SNQ KFA Jülich, Germany	Linac and compressor ring	1100	4000	100	250	^{238}U	Proposal under development
ASPUN Argonne, USA	FFAG Syncrotron	1600	4000	60	0.4	^{238}U	Proposal under development

Production of Neutrons

The *target* chosen for spallation sources must be high atomic number and high density, and in almost all cases has been uranium. At moderate intensities the neutron intensity can be increased by using fissile material in a 'booster' target. The gain factor is $G \approx 1/(1-k)$, where k is the prompt-neutron multiplication factor. Boosters have been used on electron linacs in low power levels. Plans are now almost complete for a booster target for IPNS that represents the first such target at a proton source on its installation in 1987. The IPNS booster should give a factor of about 3 in neutron-beam intensity, and for possible future applications it will be very important to measure the delayed neutron yield from the fission process and to monitor the cooling requirements of the new target. The additional heat produced in the fission process (about 190 MeV per useful neutron produced as opposed to about 50 MeV for true spallation) may make the installation of booster targets less effective at more intense sources.

Once the neutrons are produced in a target they must be slowed down to less than 1 eV in *moderators*. For pulsed sources spreading of the pulse in time must be minimised. This may involve the deliberate addition of 'poisons', or neutron absorbers, or use of low temperatures. The area of moderators represents one of the most important in terms of pulsed neutron technology. Two objectives must be satisfied simultaneously, maximum intensity and minimum pulse width for the best resolution. Moreover both of these parameters are functions of energy. To optimise each experiment may therefore call for a change in moderator design, degree of poisoning, and moderator temperature. Clearly such flexibility does not exist, but there is still room for much ingenuity in moderator design. Thin decoupled moderators produce a copious supply of under-moderated epithermal neutrons ($E>100$ meV) and, as discussed below, this has allowed some unique science to be performed. On the other hand, figure 9.1 illustrates the spectrum emerging from the solid methane cold source (physical temperature 10 K) at IPNS, The spectral temperature of this source is 20 K, and it is the coldest such source in the world. Thus, the common perception that pulsed sources are good only for high-energy

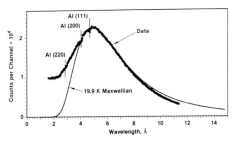

Figure 9.1 Spectral distribution from solid methane cold source at IPNS, Argonne.

neutron experiments is far too simple; small-angle scattering experiments with 14 Å neutrons are now being successfully performed at IPNS.

Pulse lengths are also an important parameter. Two general classes of accelerators and corresponding moderators are contemplated for pulsed sources. The 'short-pulse' variety is based on accelerators producing sub-microsecond pulses (IPNS, ISIS, WNR/PSR) and uses small dense decoupled hydrogeneous moderators. The 'quasi-steady' state type of pulsed source is based on accelerators which produce pulses of duration several hundred microseconds and uses large D_2O or hydrogeneous moderators, which provide efficient moderation and storage of thermal neutrons. This design has been proposed for the German SNQ† project. In the first case the pulse width $\Delta t \sim 15 \times \lambda$ μs, where λ is in Å, whereas in the second case $\Delta t \sim 500$ μs independent of energy. Time-averaged beam intensities of thermal neutrons per primary source neutron are greater for quasi-steady state sources, and they are generally optimised towards the cold end of the neutron spectrum. Wavelength separation must be performed with a subsidiary chopper or monochromator and since both of these are more efficient at low energies, this again tends to focus research efforts below 100 meV. In this region of the spectrum—conventionally the province of reactors—the quasi-steady state source brings the extra dimension of time structure. The most thorough study of the considerable advantages to be gained from time structure, ranging from the simple reduction in background when the source is 'off', to the use of phase space transformation by Doppler shifting neutrons into the same portion of phase space, has been carried out by the SNQ project scientists (Stiller 1984).

Scientific Application

A survey of instruments at pulsed sources has been given in Windsor's book (1981) and more recently by Carpenter *et al* (1984). Some of the scientific highlights of pulsed sources have been discussed in *Physics Today* (Lander and Price 1985). Space does not permit us to mention any of the numerous accomplishments in detail. There are, however, a number of themes that emerge and these are worth stating briefly.

Scattering experiments from polycrystalline samples, amorphous materials and liquids have all been very successful at spallation sources. The short wavelength neutrons, which are plentiful at these sources, allow high values of Q, the momentum transfer, to be attained and this gives additional real space resolution over that possible at reactor sources. The energy dispersive nature of the scattering, i.e. fixed angle, variable λ,

†The project Spallation Neutron Quelle, centred at Julich. In early 1986 the West German Government decided against building the SNQ in its present form, but new proposals are being prepared.

allows high pressures, temperatures, and magnetic fields, all of which normally restrict the access of the neutron beams, to be used. In small-angle scattering the use of a wide band of wavelengths allows studies to be performed from 0.004 to 0.3 Å^{-1} at a single instrument setting. This has already proved useful at IPNS and KENS for the investigation of polymers and formation of precipitates in materials science. When the lower end of the range is extended to 0.001 Å^{-1} or less, these instruments will be unique.

The epithermal ($E_0 > 100$ meV) beams from these sources allow relatively large values of the energy transfer at small momentum transfer to be attained. This is particularly useful in vibrational spectroscopy, magnetism and with amorphous materials, and completely new fields have been opened by the chopper spectrometers at IPNS and the Harwell linac. Another area that has also been dominated by the chopper spectrometers, especially those at IPNS, has been the determination of momentum distributions $n(p)$. These are experiments also using high incident energies, but measurements are performed at high values of Q to ensure that the impulse approximation is valid. They are experiments of particular importance in the quantum solids H, ^{3}He and ^{4}He and give unique information on the many-body wavefunctions.

Future Accelerators

The first source using a proton accelerator was ZING-P at Argonne and had an energy of 200 MeV and a time-averaged current of 0.1 μA. The ASPUN or SNQ projects now advanced at Argonne and Jülich, respectively, have energies of about 1600 MeV and projected currents of 4000 μA. These represent an increase in neutron intensity from the early ZING-P source by a factor of $\sim 10^6$. Figure 9.2 shows this variation plotted as a function of time, normalised to IPNS in 1984. The almost logarithmic increase of a factor of 10 about every 5 years contrasts sharply with the very small gains in reactor fluxes—the world's highest flux reactor HFIR was commissioned at Oak Ridge in 1966—and is an impressive testimony to accelerator developments.

To produce submicrosecond-pulse proton beams the common method is the use of a synchrotron machine. However, the most advanced of such machines have a design goal of accelerating about 6×10^{13} protons per pulse (ppp). IPNS runs at 3×10^{12} ppp and the Rutherford ISIS at full power (800 MeV, 200 μA) will have 2.5×10^{13} ppp. Scientists at KEK, Tsukuba, Japan, have discussed a synchrotron (GEMINI) that could possibly deliver 500 μA at 800 MeV, which would have a space-charge limit of about 7×10^{13} ppp and clearly be at the forefront of accelerator technology. To obtain higher currents one must either use linacs, or fixed

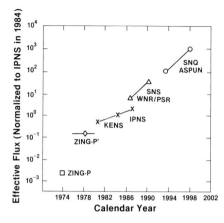

Figure 9.2 Plot of available neutron intensity from spallation sources from 1974 to 2002. The continuous bars represent the fact that after accelerators are turned on they usually take some years before they reach full potential. Thus the experimentalist can expect about a factor of 5 to 10 increase in intensity during the first 10 years of operation.

field alternating gradient (FFAG) type machines. Linacs produce long pulses (LAMPF at Los Alamos, for example, has a pulse width of $750\,\mu s$) and may require compressor rings. The SNQ design specifies a pulse length of $250\,\mu s$ with plans to build a compressor ring at a later stage.

At Argonne the ASPUN project, developed by Khoe and Kustom (1984), relies on the FFAG concept. In this design the proton orbits increase in radius as the energy increases. The magnetic field also increases with radius to provide increased bending strength. However, these fields do not vary with time and may also use superconducting magnet technology. The direct-current fields allow more efficient injection and capture of the beam and more effective use of the radio-frequency systems. The FFAG synchrotron also allows the opportunity to stack beams inside the machine, thus having an extracted repetition rate different from that at injection. Work continues on the accelerator design of both SNQ and ASPUN and no doubt new ideas will emerge.

Future Developments

Pulsed neutrons are relatively new. As with many new scientific technologies the instrumentation for them is developing at a rapid rate and it is too early to predict the scientific rewards. We mention here briefly four new areas:

(1) Spectroscopy in the eV region. Instruments have been set up at a number of laboratories, based on using sharp resonances in materials such as ^{238}U, which has a resonance at 6.67 eV. Identification of the final neutron energy is done either indirectly by neutron capture or by

detection of the secondary particles (γ rays from ^{238}U) emitted. Experiments are aimed at either very high-Q recoil scattering ($Q \sim 100$ Å^{-1}) or electronic excitations at low Q but large energy transfer.

(2) Spectrometers to examine modes with dispersion in single crystals. This area of investigation has been a most important one for neutrons and two types of instrument designed to study such excitations are under development at pulsed sources. Experience at both Harwell, with the electron linac, and at the Japanese source KENS has shown that modes with dispersion can be measured successfully along symmetry directions at a pulsed source. An instrument of this type is also being built at Los Alamos.

(3) Small-angle scattering and the use of resonance isotopes. New experiments designed to capitalise on the dynamic range in Q mentioned earlier are currently being explored. At IPNS experiments have recently been performed with chelating agents designed to entrap metal ions and have shown that the use of resonance effects with ^{157}Gd (resonance wavelength=1.6 Å) greatly increases the sensitivity of the neutron results to the number and form of the Gd atoms entrapped in the micelles. This is quite new application of neutron small-angle scattering with potential applications in biology.

(4) Time-resolved studies. Clearly, the pulsing nature of the source can be used to advantage if the stimulus to the sample can be phased with the arrival of the neutron burst. Pioneering experiments including time resolution were performed at the Tohoku electron linac (Niimura and Muto 1973) to examine the motion of the domains stimulated by an electric field in the ferroelectric material $NaNO_2$. More recently, again in Japan, relaxation phenomena in spin glasses have been the subject of time-resolved neutron experiments. These indicate a considerable potential for studies in the time domain of about 1 μs to 10 ms, and a number of investigations are under active development at pulsed neutron facilities. We should note here that since a 1 Å neutron travels at a velocity of 4 mm μs^{-1}, studies in the time domain less than 1 μs appear unlikely with present or planned neutron sources and samples.

We cannot end without once again pointing out the importance of future sources, and instrumentation, for pulsed neutrons. All sources currently operating, or even under construction, rely on accelerator concepts which originated for other scientific applications in high-energy or nuclear physics. Certainly, major investments have been made in facilities such as the Rutherford ISIS, and the Los Alamos WNR/PSR which should move us forward in a number of areas. Nevertheless, accelerator physicists are now beginning to design new types of machine specifically for pulsed neutron research, and the results such as the German SNQ or the Argonne ASPUN proposal are formidable machines. They will represent state-of-the-art proton accelerators in terms of proton current, although the energy requirements are rather modest at <2 GeV. The production of such proton beams will be of interest to a large number of scientists interested in nuclear physics, neutrino production and medical applications in addition to condensed matter research. They are also very expensive. Building sources of this kind will require the fullest degree of cooperation at the national, if not international, level.

References

Carpenter J M 1977 *Nucl. Instrum. Methods* **145** 91
Carpenter J M, Lander G H and Windsor C G 1984 *Rev. Sci. Instrum.* **55** 1019
Khoe T K and Kustom R L 1984 *IEEE Trans. Nucl. Sci.* **NS 30** 2086
Lander G H and Price D L 1985 *Phys. Today (January)* 38
Niimura N and Muto M 1973 *J. Phys. Soc. Japan* **35** 628
Stiller H (ed) 1984 *SNQ Information, No 5* KFA, Jülich, W. Germany
Windsor C G 1981 *Pulsed Neutron Scattering* (London: Taylor and Francis)

9.2   ISIS—The UK Pulsed Spallation Neutron Source

G C Stirling

Rutherford Appleton Laboratory, UK

The spallation neutron source at the Rutherford Appleton Laboratory is a pulsed neutron facility designed primarily for thermal neutron scattering research. Its successful first operation in December 1984 represented a notable step in the world-wide campaign by materials scientists for more intense neutron beams, and the facility promises to be the state-of-the-art pulsed neutron source into the next decade.

Technical descriptions have been well documented elsewhere, and will not be repeated here. Likewise experimental results, which above all else give a realistic measure of performance, may be read in the scientific literature. This account, instead, concentrates on the background leading

Figure 9.3 Ceremony to mark the inauguration of ISIS at the Rutherford Appleton Laboratory on 1 October 1985 (left to right: Dr G Manning, Professor E W J Mitchell, The Prime Minister, Sir John Kingman).

to the adoption of the project, with particular emphasis on the role played by research scientists from UK universities and elsewhere.

The spallation neutron source was formally inaugurated by Margaret Thatcher, Prime Minister of the United Kingdom, on 1 October 1985, and named ISIS†. The ceremony (figure 9.3) coincided with the commencement of the scientific research programme, and confirmed the importance of neutron beams for research into the structure and properties of materials.

Historical Background

Neutron scattering research, in Britain as elsewhere, grew out of the atomic power programme, and in the 1950s and 1960s was mainly based at the Atomic Energy Research Establishment at Harwell. During this period a continually increasing number of university scientists used the Harwell

---

†This name now replaces the acronym SNS which has been widely used and appears in the literature.

facilities, especially the DIDO and PLUTO reactors, in their research. A significant feature of this growth was the widening range of application of neutron techniques, from solid-state and liquid physics to chemistry, in particular, and biology.

By the late 1960s the majority of basic neutron scattering research in the UK was being carried out by university scientists, supported by the Science Research Council, an arm of the Department of Education and Science. It was rapidly becoming obvious that access to higher intensity beams was essential if demands for beam time by the growing community were to be satisfied. There was also a need to keep up with developments elsewhere, in particular the high-flux reactors which were already operational in the United States and the Institut Laue–Langevin (ILL) reactor under construction at Grenoble, France.

The university neutron scattering programme was directed then, as now, by the Neutron Beam Research Committee (NBRC) of the Science Research Council, a body with a rotating membership and largely consisting of university scientists. Most, though not all, have an involvement in neutron scattering research. In 1971 it was resolved that a permanent unit should be established in order to carry out the Committee's capital projects, and this would be based at the Science Research Council's Rutherford Laboratory next-door to Harwell. More importantly in the present context, the unit was charged with formulating plans to meet the future growth of the subject. The unit worked in close cooperation with the NBRC (and has continued to do so) and acted as the focus for plans to develop a new neutron source. Attention was immediately given to the proposal to build a high-flux reactor in the UK, but this was eventually dropped in 1972 in favour of a decision to join the Franco-German Institut Laue–Langevin in Grenoble, Britain becoming an equal third partner. The availability of additional facilities at ILL had an immediate and dramatic effect on the UK neutron scattering programme, shown in figure 9.4. The growth in the numbers of UK users closely matched the provision of new instruments at ILL, only levelling off when the first phase of instrument construction was complete and the beam time became heavily oversubscribed. The conclusion was drawn that facilities were still inadequate to support the potential demand, consistent with the earlier judgement that the UK programme required access to a completely new source of neutrons.

In anticipation of this unfulfilled demand, the NBRC had, in 1973, adopted an outline study programme directed towards the provision of 'a next generation source in the UK by about 1985'. The programme envisaged a review of possible candidates followed by feasibility and design studies, which would lead to a firm proposal by the late 1970s. Attention was initially given to four main categories, namely steady state reactors, pulsed reactors, accelerator-based systems, with or without neutron boos-

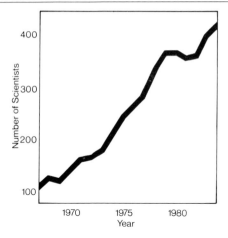

Figure 9.4 Growth in UK neutron-scattering community 1967–84.

ters, and fusion systems. An important consideration concerned the distinction between steady state and pulsed sources. At that time there was little experience in the utilisation of pulsed neutron techniques, but it had been speculated 'that even if an order of magnitude improvement in the present reactors were feasible, the ensuing science is likely to follow well-trodden paths, whereas a pulsed source may provide the more exciting possibilities for exploring new fields'. The veracity of this observation is still a subject for spirited debate; notwithstanding, the decision was made in 1975 to concentrate effort on pulsed systems.

At that time British experience with pulsed sources was based largely on work carried out at the Harwell 30 MeV electron linac source which operated with a mean power of 4 kW. Although relatively modest, the Harwell linac had proved to be competitive with existing steady-state sources for some applications, notably in the short-wavelength region. In 1975 Harwell received approval to replace this machine with a new electron linac, which would give an order of magnitude increase in neutron intensities. Although funded for nuclear physics use, multiplexing for condensed matter research was envisaged, and planning for eventual utilisation was actively being pursued by Harwell and Science Research Council teams.

At the same time, interest in pulsed sources was growing world-wide, in particular those based on pulsed spallation systems. In the United States, prototype experiments at the Argonne National Laboratory had led to the IPNS proposals, and the WNR facility was under construction at Los Alamos; in Japan there were proposals to build a neutron facility using the new proton synchrotron at KEK. All these proposals stemmed from the same factors: first, the growing realisation that pulsed sources provided the only realistic way to attain higher effective intensities than those currently available; second, the efficiency of the proton spallation reaction, whereby copious fluxes of neutrons can be produced at relatively low heat output per neutron produced; and third, the availability of existing accelerators which could be used either for prototype experiments or, in a parasitic mode, for a fully fledged facility. While the attractions of proton spallation were becoming clearer, a concurrent design study at Oak Ridge had equally shown that substantial performance could be achieved with a neutron booster installed on an electron linac. The NBRC consequently decided to include the option of a booster on the new Harwell linac in the list of candidates for a new neutron source.

At the Rutherford Laboratory, it was realised that valuable measurements on spallation neutron production could be performed on the ageing 7 GeV proton synchrotron Nimrod, and a programme to study target and moderator performance using Nimrod proton beams was drawn up. At that time plans had been prepared to replace Nimrod with a new accelerator complex for high-energy particle physics named EPIC, but late in 1975 these were dropped following the go-ahead for an equivalent scheme in Germany. It was immediately recognised that the potential availability of Nimrod plant and equipment might provide a substantial base for a new proton spallation neutron source. The concept was enthusiastically endorsed and outline designs drawn up. The basic specification envisaged a new 800 MeV proton synchrotron, fed by the existing Nimrod 70 MeV injector linac, and operating at 50 Hz. Neutrons would be produced in a uranium target in an existing experimental hall. It was clear that this would have a substantially superior performance to any other facility yet approved, and that the availability of existing buildings, accelerator equipment and general infrastructure, would mean that the new source could be built for a fraction of the cost of a new construction on a green-field site.

Plans for the new facility were widely discussed by the neutron-scattering community. Some 60 scientists, representing over 20 different university departments, worked in four main working groups—solid-state physics, fluids and amorphous solids, structure determination, and molecular and biological sciences—to examine the scientific basis for the new scheme. Particular attention was given to the areas where the pulsed source would provide unique capability with opportunities for exploring new areas of science, and to the complementarity of the new source with existing facilities. In due course a formal proposal, including the reports of the working groups, was adopted by the Neutron Beam Research Committee, and government approval to proceed was granted in June 1977. An immediate decision was taken to form a science planning group to be concerned with utilisation of the new source. The group would play an advisory role to the project team at the Laboratory, and membership was drawn from a broad spectrum of the user community. An important feature was the inclusion of overseas representation, reflecting the strongly held view that the facility should be open to international participation.

Detailed design work on the accelerator, the target station and neutron scattering instruments continued during 1977–8, and manufacture and installation of equipment commenced after the final Nimrod shutdown in June 1978. Subsequent milestones in the construction of the facility are given in table 9.1 and the basic specification in table 9.2.

Short pulses of protons from the high intensity synchrotron are fed to a compact uranium-238 target located in a massive shield. The target is surrounded by four independent moderators to give a range of neutron beam characteristics, both in energy (wavelength) spectra and time structure: ambient and epithermal neutron beams are produced from poisoned H_2 moderators, high resolution beams from a 100 K liquid CH_4 moderator, and cold neutrons from a 25 K liquid H_2 moderator. There are 18 beam ports, giving the capability for some twenty or more instruments which will have a wide range of applications to cater for the diverse community of users. There will be a continuing programme of instrument provision.

Figure 9.5 is a general external view of the whole installation.

Table 9.1 ISIS milestones.

June 1977	Project approval
June 1978	Shutdown of Nimrod
November 1981	First magnets installed in synchrotron
December 1983	Synchrotron ring closed
January 1984	70 MeV beam circulated in synchrotron
June 1984	Beam accelerated to 550 MeV
December 1984	Beam transported to target station; first neutrons
October 1985	ISIS inauguration; science programme underway

Table 9.2 ISIS parameters.

Proton injection energy	70 MeV
Final proton energy	800 MeV
Proton pulse length	$0.4\,\mu$s
Pulse frequency	50 Hz
Proton intensity	2.5×10^{13} protons per pulse
Average proton current	$200\,\mu$A
Target	Depleted uranium-238
Moderators	316 K, H_2O
	100 K, CH_4
	21 K, H_2

Figure 9.5 The spallation neutron source ISIS. The injector and synchrotron ring are seen top centre, and the large experimental hall lower left. Note the 90 m neutron guide to the right of the experimental hall, serving the high resolution powder diffractometer.

Source Performance

ISIS was operated as a whole for the first time on 16 December 1984, with the aim of checking *in toto* the performance of the accelerator, the target and moderator assemblies, and the neutron scattering instruments with their attendant data acquisition systems. Even at the deliberately low beam intensity of this first demonstration run, it proved possible to make new diffraction measurements which previously had been impossible on existing neutron sources. A notable example was the observation of the magnetostrictive distortion in nickel oxide using the high resolution powder diffractometer (HRPD). This is manifested by splitting of the 111 and $\overline{1}11$ peaks in the powder pattern, and made possible by the very high resolution given by the time-of-flight technique.

Commissioning of HRPD during 1985 quickly established it as the most powerful instrument of its type in the world, the resolution being some 5–7 times better than existing neutron powder diffractometers. As well as allowing more complex problems of a traditional nature to be tackled, such as structure refinement using the Rietveld method, the increased resolution dramatically reduces the extent of peak overlap and opens up new scientific possibilities such as detailed line-broadening studies, the detection of subtle symmetry changes and, of particular importance, the potential to solve unknown crystal structures in an *ab initio* manner using autoindexing and direct methods techniques. The feasibility of this latter

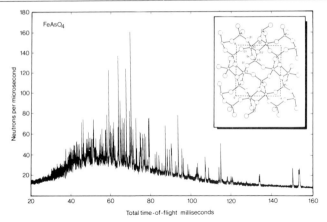

Figure 9.6 The time-of-flight neutron powder diffraction of $FeAsO_4$ prior to normalisation. Inset: the (001) projection of the crystal structure of $FeAsO_4$ determined by auto-indexing and single-crystal direct methods. The complex monoclinic structure was obtained using triplets and negative quartets in the direct methods package MITHRIL and consists of tetrahedrally coordinated arsenic and edge-shared 5-coordinated iron ions.

procedure was demonstrated at only 0.5% full intensity with the determination of the unknown structure of $FeAsO_4$ for which single crystals are not available. Data are shown in figure 9.6. The structural determination took place in four discrete steps: (*a*) automatic indexing of the non-overlapped diffraction peaks in the powder pattern and the determination of the crystal system and lattice parameters, (*b*) identification of the possible space group from systematic absences, (*c*) solution of the phase problem and determination of an approximate structure by direct methods, (*d*) refinement of the structure.

When the scientific programme started in 1985, nine operating instruments were available, including powder and single-crystal diffractometers, inelastic and quasielastic scattering spectrometers with different resolutions and momentum and energy transfer ranges, and polarised neutrons.

Future Developments

In 1985 the main objective is to run the source up to its full design performance. Proton currents will be increased as experience is gained on the behaviour of the accelerator systems. An important constraint in this work is the necessity to avoid undue beam loss, with the attendant activation of components. Concurrently, the suite of neutron scattering instruments will be built up, with close attention being given to the evolving requirements of the university user groups. The trend towards internationalisation will lead to wider opportunities for exploitation. Already neutron instruments are being provided by overseas groups. Another example is given by the muon facility which has been jointly funded by a consortium of European countries under the umbrella of the European Commission. This will use a small fraction of the proton beam to produce polarised muon beams for condensed matter research using the μSR technique. It will operate independently of the neutron programme, but there will undoubtedly be a healthy overlap between the user communities. Yet a different example of non-neutron exploitation and international collaboration is the neutrino facility KARMEN, a joint venture between the UK and the Federal Republic of Germany to be operational in 1987. Meanwhile studies are in progress to ensure that the source maintains a role as a world-leader into the future. Options include improvements to the accelerator systems, provision of booster target assemblies, and the provision of new and improved instrumentation. It is expected that much of this work will benefit from international collaboration.

Conclusions

At first sight, British scientists in 1985 had enviable opportunities for pursuing neutron beam research: a solid grounding in the technique derived in part from the early days at Harwell; a third share in the ILL high-flux reactor, by far the world's most productive facility; and a new facility, ISIS, poised to become the world's premier pulsed source. How had this position been reached and what would be the prospects for the future?

An immediate answer is given by comparisons with other countries having similarly strong traditions, but which now find themselves falling behind in days of limited funding. First, the necessity for scientists to speak with a coherent voice can be vital. In the United Kingdon, the establishment in 1966 of the Neutron Beam Research Committee, representing all university scientists interested in neutron scattering, and through which all funding was channelled, was of crucial importance. The subsequent formation of a neutron beam research unit enabled the Committee to draw up long-range plans and eventually, when redundant plant became fortuitously available at the Rutherford Laboratory, to carry through the spallation neutron source project.

Opportunities in the future may not be so certain. An immediate aim is to raise ISIS to its design specification, with a full set of instruments; further developments to increase the performance by substantial factors can already be foreseen. The limiting factor will undoubtedly be the level of funding, and it seems certain that this may only be achieved by increasing international collaboration. In that event, the prospects for all participants would be bright indeed.

9.3   A US Proposal for the 1990s—The Center for Neutron Research: Oak Ridge

R M Moon

Oak Ridge National Laboratory, USA

In the United States there was a fairly regular progression of federally funded research reactors constructed during the 1940s, 1950s and early 1960s. This progression showed an overall gain in flux of three orders of magnitude in going from the X-10 graphite reactor at Oak Ridge to the high-flux reactors at Brookhaven and Oak Ridge. This progress stopped abruptly with the completion of the National Bureau of Standards Reactor in 1967; the newest national US reactor is now almost twenty years old and will be about thirty years old before another major research reactor can be constructed.

For the past ten years, the US scientific community, through a series of committees formed at the request of various government agencies, has repeatedly called for the beginning of design efforts on a new neutron source of higher flux. In order to get this idea out of the talking stage and into the action stage, the Oak Ridge National Laboratory has begun an effort to design a new reactor capable of achieving a flux of at least 5×10^{15} neutrons cm^{-2} s^{-1} in a large D_2O reflector.

The primary motivation is to build unsurpassed neutron scattering facilities, but the reactor will also be used for isotope production and materials irradiation experiments. The conflicting requirements of these three major programmes present a challenge to the reactor designers. Neutron scattering needs high thermal flux in the reflector. For the production of certain important transuranic isotopes, a high epithermal flux is desired; for other isotopes, a high thermal flux is necessary. Materials irradiation studies require a high fast neutron flux. A general concept with annular fuel, similar to that of HFIR, with D_2O as coolant, moderator, and reflector shows great promise in meeting all these requirements. A preliminary calculation of flux as a function of radial distance is shown in figure 9.7. Isotope production facilities would be located in the central target region (high epithermal flux) and in the reflector (high thermal flux). Instrumented materials irradiation facilities would be located as close to the fuel as possible (high fast flux). The opportunity to incorporate all the advances in neutron scattering technology is at least as important as the

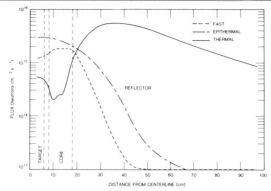

Figure 9.7 Preliminary calculation of the radial variation of neutron flux for various energy groups for a 200 MW reactor with heavy water as coolant, moderator and reflector.

higher source flux. Beam tubes will be designed to take full advantage of vertically focusing monochromators. There will be two cold sources and at least eight cold beams delivered to a large guide hall. Compared to the HFIR, the useful flux at the sample position will be at least 10 times greater for most experiments. Current plans call for about 30 neutron beam experiments.

To reach a flux of 5×10^{15} neutrons cm^{-2} s^{-1} the power density will have to be increased substantially over that of current reactors. The preliminary plans call for operating at an average power density of 5 MW l^{-1} and a total power of 200 MW. The limiting factor for power density in Al-clad fuel is the growth of an oxide layer on the surface of the cladding, resulting in a large temperature change across this layer for high heat fluxes. At present, the HFIR operates with a power density of 2 MW l^{-1}; the limit with

Figure 9.8 R M Moon and W C Koehler at the Gatlinburg Conference in June 1976.

existing fuel technology is probably 3–4 MW l^{-1}. Surface modification techniques to inhibit the growth of the oxide layer will be explored in the reactor development programme.

The reactor and associated experimental and office facilities, which have been named the Center for Neutron Research (CNR), will be open to use by scientists of many disciplines from universities, industries and other government laboratories. In many respects it will be similar to the Institut Laue–Langevin, but with a flux about four times higher. No official decision has been made to construct the CNR; under the present schedule construction would not begin until 1990 with completion in 1995.

9.4 Neutron Scattering in the Future— A Personal View

J E Enderby

Institut Laue–Langevin, Grenoble, France

Neutron scattering is a uniquely powerful tool for the study of the structure, the elementary excitations and phase transitions of condensed matter (glasses, crystalline solids and liquids). This is because thermal neutrons are the only probe for which the energy and wavevector are both well matched to the relevant scales of energy and distance.

The pioneering work of Brockhouse on the determination of phonon and magnon dispersion relations, for example, is now described in solid state physics textbooks and underpins a great deal of our detailed know-ledge of interatomic forces and of exchange interactions. This work led to the development of the shell model and its various extensions to describe the forces in ionic materials and semiconductors, and provided essential support for the pseudo-potential theory of metals.

In the future, work will be concentrated on novel types of excitations where the atomic motions are more complex than those associated with small vibrations of regular crystals. One example concerns incommensur-ately modulated phases where the relevant excitation is called a 'phason'. To study this type of excitation, new instruments with higher resolution will be required, and in the next ten years improvements in collimators and monochromators and in detection techniques will allow much wider sur-veys to be undertaken of both reciprocal space and energy.

Back-scattering and spin echo techniques will also be further developed so that extra high resolution quasi-elastic scattering experiments will become routine. This will have important consequences for the study of dynamic processes in polymers, particularly those involving long time-scales. Small-angle scattering will also continue to be improved and will have increasing industrial application. The possibility of doing real-time experiments in which periodic external forces are applied to samples will be fully realised.

Emphasis is likely to be placed on structural changes as a function of some external variable such as temperature, pressure, magnetic field or chemical environment, and increased rates of data collection will permit *in situ* studies of such changes.

There will still be a need for the location of hydrogen atoms, but the tendency will be to study very large systems such as metal-coordination complexes and biological macromolecules. There is stong interest in precision studies at high resolution and all of these rely on improvements in measurement techniques (notably multidectors) and computer power. For liquids and amorphous solids, the techniques of isotopic substitution based on first- and second-order difference methods will be further developed and will be tried on isotope pairs where the scattering amplitude difference is small. Perhaps the most exciting example is ^{12}C and ^{13}C where the small difference in scattering length could, with more intense sources, be exploited successfully in sorting out the local structure around that most ubiquitous element—carbon.

There will be an increased interest in the physics of the neutron itself. This will arise from the need to provide sharp tests of the fundamental theories of physics (Grand Unification Theories, supersymmetry etc) and the increasing costs (perhaps even impossibility) of achieving the necessary energy by 'conventional' acceleration methods. The n–n̄ oscillation experiment, planned at the ILL in the next year or so, is one such example. Ultra cold neutrons will probably find increased use in this domain as well as in surface physics.

Perhaps the best indication of the health and vitality of neutron scattering in the 1990s is the investment in new or improved sources and instrumentation which will be made in the remainder of the 1980s and early 1990s. At the Institut Laue–Langevin, for example, the original cold source has been replaced with one of improved design to enhance the cold neutron flux in the existing guides. In addition, a second cold source is under commission, allowing an increased number of cold neutron instruments to be installed. Possibilities for increasing the reactor flux are also being considered. In Germany and in North America, considerable improvements are now in progress with more in the planning stage. In the UK a major investment in pulsed source technology, particularly at the Rutherford–Appleton Laboratory, is now coming to fruition (ISIS).

The final prediction is that neutron experimenters will become increasingly organised on a world-wide scale and will learn how to exploit the complementary nature of steady state sources like those located at the ILL and ORNL and pulsed sources such as ISIS (RAL) and IPNS (Argonne). The power of the neutron method has still not been fully appreciated by the wider scientific community, primarily, in my view, because of the lack of beam time and the feeling that the technique itself is very specialised. This situation will surely change as we go into the 1990s.

Bibliography

Bacon G E 1955 *Neutron Diffraction* (Oxford: Clarendon) 1st edn (3rd edn 1975)
(Russian transl. 1st edn published in Moscow, 1957, Chinese transl. 3rd edn
published in Beijing, 1980)
——1966 *X-ray and Neutron Diffraction* (with a collection of early original papers)
(Oxford: Pergamon)
Boutin H and Yip S 1968 *Molecular Spectroscopy with Neutrons* (Boston, MA:
MIT Press)
Convert P and Forsyth J B (eds) 1983 *Position-Sensitive Detection of Thermal
Neutrons* (London: Academic)
Curtiss L F 1959 *Introduction to Neutron Physics* (New York: Van Nostrand)
Dachs H (ed) 1978 *Neutron Diffraction* (Berlin: Springer)
Dorner B 1982 *Coherent Inelastic Neutron Scattering in Lattice Dynamics* (Berlin:
Springer)
Egelstaff P A 1965 *Thermal Neutron Scattering* (London: Academic)
Faber J Jr (ed) 1981 *Neutron Scattering Symposium at Argonne* AIP Conference
Proceedings No. 89
Gurevich I and Tarasov I V 1968 *Low Energy Neutron Physics* (Amsterdam:
North-Holland) (translated from the Russian, Moscow 1965)
Hughes D J 1954 *Neutron Optics* (New York: Interscience)
International Atomic Energy Agency 1960 *Inelastic Scattering of Neutrons in Solids
and Liquids, Symposium at Vienna, 1960* STI/PUB/35 (Vienna: IAEA)
—— 1960 *Pile Neutron Research in Physics, Symposium at Vienna, 1960* STI/PUB/
36 (Vienna: IAEA)
—— 1962 *Inelastic Scattering of Neutrons in Solids and Liquids, Symposium at
Chalk River, 1962* Vols I, II STI/PUB/62 (Vienna: IAEA)
—— 1964 *Inelastic Scattering of Neutrons, Symposium at Bombay, 1964* Vols I, II
STI/PUB/92 (Vienna: IAEA)
—— 1968 *Neutron Inelastic Scattering, Symposium at Copenhagen, 1968* Vols I, II
STI/PUB/187 (Vienna: IAEA)
—— 1972 *Neutron Inelastic Scattering, Symposium at Grenoble, 1972* STI/PUB/308
(Vienna: IAEA)
—— 1977 *Neutron Inelastic Scattering, Symposium at Vienna, 1977* Vols I, II
STI/PUB/468 (Vienna: IAEA)
Izumov Yu A and Ozerov R P 1970 *Magnetic Neutron Diffraction* (New York:
Plenum) (translated from the Russian, first published by Nauka Press, Moscow,
1966)
Koester L and Steyerl A 1977 *Neutron Physics* (Berlin: Springer)
Lander G H and Robinson R A (eds) 1986 *Neutron Scattering* (Amsterdam:
North-Holland)
Larose A and Vanderwal J 1974 *Scattering of Thermal Neutrons, a bibliography of
published papers (1932–74)* (New York: Plenum)
Lovesey S W 1984 *The Theory of Neutron Scattering from Condensed Matter* Vol 1.
Neutron Scattering. Vol 2. *Polarization Effects and Magnetic Scattering* (Oxford:
Oxford University Press)

Lovesey S W and Springer T (eds) 1977 *Dynamics of Solids and Liquids by Neutron Scattering* (Berlin: Springer)

Marshall W and Lovesey S W 1971 *Theory of Thermal Neutron Scattering* (Oxford: Clarendon)

Moon R M (ed) 1976 *Proc. Conf. on Neutron Scattering at Gatlinburg, 1976* Vols I, II, Oak Ridge National Laboratory, CONF-760601-P1, 2

Schoenborn B P (ed) 1979 *Neutron Scattering for the Analysis of Biological Structures, Symposium at Brookhaven* BNL 50453

Schofield P (ed) 1982 *The Neutron and its Applications* (Inst. Phys. Conf. Ser. 64)

Squires G L 1978 *Introduction to the Theory of Thermal Neutron Scattering* (Cambridge: Cambridge University Press)

Stevenson R W H (ed) 1966 *Phonons in Perfect Lattices and in Lattices with Point Imperfections* (Edinburgh: Oliver & Boyd)

Willis B T M (ed) 1970 *Thermal Neutron Diffraction* (Oxford: Oxford University Press)

—— (ed) 1973 *Chemical Applications of Thermal Neutron Scattering* (Oxford: Oxford University Press

Windsor C G 1981 *Pulsed Neutron Diffraction* (London: Taylor and Francis)

Subject Index

Author Index